Safety and
Human Error in
Engineering Systems

Safety and Human Error in Engineering Systems

B. S. Dhillon

CRC Press
Taylor & Francis Group
Boca Raton London New York

CRC Press is an imprint of the
Taylor & Francis Group, an **informa** business

CRC Press
Taylor & Francis Group
6000 Broken Sound Parkway NW, Suite 300
Boca Raton, FL 33487-2742

First issued in paperback 2019

© 2013 by Taylor & Francis Group, LLC
CRC Press is an imprint of Taylor & Francis Group, an Informa business

No claim to original U.S. Government works

ISBN-13: 978-1-4665-0692-3 (hbk)
ISBN-13: 978-0-367-38115-8 (pbk)

Library of Congress Cataloging-in-Publication Data

Dhillon, B. S. (Balbir S.), 1947-
    Safety and human error in engineering systems / B.S. Dhillon.
       p. cm.
    Summary: "This book combines coverage of safety and human factors in engineering into a single volume, eliminating the need to consult many different and diverse sources to obtain desired information. It discusses mathematical concepts necessary to understanding material presented in subsequent chapters. The author describes the methods that can be used to perform various types of safety and human error analysis in engineering systems. The book contains examples along with their solutions and at the end of each chapter are numerous problems to test reader comprehension"-- Provided by publisher.
    Includes bibliographical references and index.
    ISBN 978-1-4665-0692-3 (hardback)
    1. Industrial safety. 2. Errors--Prevention. 3. Human engineering. I. Title.

T55.D485 2012
363.1--dc23   2012004789

Visit the Taylor & Francis Web site at
http://www.taylorandfrancis.com

and the CRC Press Web site at
http://www.crcpress.com

This book is affectionately dedicated to all my friends including Matt Farrow, Joe Hazelton, and Archy during my teenage years in England for their kindness and guidance.

Contents

Preface

Nowadays, engineering systems are an important element of the world economy because each year billions of dollars are spent to develop, manufacture, and operate various types of engineering systems around the globe. Their safety and failure due to human error have become an important concern because of increasing accidental deaths and cost. For example, in regard to automobile accidents on highways alone, in the United States around 42,000 deaths occur annually and, in 1994, the cost of motor vehicle crashes was estimated to be about $150 billion to the United States economy.

Furthermore, around 70–90% of transportation-related crashes were directly or indirectly due to human error. Needless to say, safety and human error in engineering systems have become more important than ever before. Over the years, a large number of journal and conference proceedings articles on various aspects of safety and human error in engineering systems have appeared, but to the best of this author's knowledge there is no specific book on the topic. This causes a great deal of difficulty for information seekers because they have to consult many different and diverse sources.

Thus, the main objective of this book is to combine safety and human error in regard to engineering systems into a single volume and to eliminate the need to consult many different and diverse sources in obtaining desired information. The book contains a chapter on mathematical concepts considered necessary to understand material presented in subsequent chapters.

The topics covered in the book are treated in such a manner that the reader will require no previous knowledge to understand the contents. At appropriate places, the book contains examples along with their solutions, and at the end of each chapter there are numerous problems to test the reader's comprehension. The sources of most of the materials presented are given in the reference section at the end of each chapter. An extensive list of publications dating from 1926 to 2009, directly or indirectly on safety and human error in engineering systems, is provided at the end of this book to give readers a view of the intensity of developments in the area.

This book is composed of 11 chapters. Chapter 1 presents various introductory aspects of safety and human error including safety and human error-related facts and figures, terms and definitions, and sources for obtaining useful information on safety and human error in engineering systems. Chapter 2 reviews mathematical concepts considered useful to understanding subsequent chapters. Some of the topics covered in the chapter are Boolean algebra laws, probability properties, probability distributions, and useful definitions.

Chapter 3 presents various introductory safety and human factors and error concepts. Chapter 4 presents a total of 10 methods considered useful for performing safety and human error analysis in engineering systems. These methods are interface safety analysis, technic of operations review, root cause analysis, hazards and operability analysis, preliminary hazard analysis, failure modes and effect analysis (FMEA), probability tree method, error-cause removal program, Markov method, and fault tree analysis. Chapter 5 is devoted to transportation systems safety. Some of the topics covered in this chapter are railroad tank car safety, light-rail transit-system safety issues, truck and bus safety-related issues, causes of airplane crashes, and ship port-related hazards.

Chapters 6 and 7 present various important aspects of medical systems safety and mining equipment safety, respectively. Chapter 8 is devoted to robot and software safety. It covers topics such as safety considerations in robot life cycle, human factors issues in robotic safety, general guidelines to reduce robot safety problems, software hazard causing ways, basic software system safety-related tasks, and software hazard analysis methods. Chapter 9 covers various important aspects of human error in transportation systems including railway personnel tasks prone to serious human error, typical human error occurrence areas in railway operation, common driver errors and ranking of driver errors, contributory factors to flight crew decision errors, and shipping systems human error-related facts and figures.

Chapter 10 is devoted to human error in healthcare systems and in mining equipment. Some of the topics covered in the chapter are healthcare systems human error-related facts and figures, medical device operator errors and medical devices with a high incidence of human errors, general guidelines for reducing medical device/equipment user interface-related errors, causes and classifications of human errors resulting in fatal mine accidents, common mining equipment-related maintenance errors and their contributory factors, and methods to perform mining equipment human error analysis. Finally, Chapter 11 presents various important aspects of human error in power plant maintenance and aviation maintenance.

The book will be useful to many individuals including system engineers, design engineers, human factors engineers, safety engineers, engineering managers and administrators, researchers and instructors involved with engineering systems, and graduate and senior undergraduate students in system engineering, human factors engineering, safety, and psychology.

The author is deeply indebted to many individuals, including family members, friends, colleagues, and students for their invisible input. The unseen contributions of my children also are appreciated. Last, but not least, I thank my wife, Rosy, my other half and friend, for typing this entire book and for her timely help in proofreading.

B. S. Dhillon

About the Author

B. S. Dhillon, PhD, is a professor of Engineering Management in the Department of Mechanical Engineering at the University of Ottawa, Ontario, Canada. He has served as a chairman/director of the Mechanical Engineering Department/Engineering Management Program for over 10 years at the same institution. Dr. Dhillon is the founder of the probability distribution called *Dhillon Distribution* used by statistical researchers in their publications around the world. He has published over 362 (i.e., 215 journal and 147 conference proceedings) articles on reliability engineering, maintainability, safety, engineering management, etc. He is or has been on the editorial boards of 11 international scientific journals. In addition, Dr. Dhillon has written 39 books on various aspects of reliability, safety, healthcare, engineering management, design, and quality that were published by John Wiley & Sons (1981), Van Nostrand (1982), Butterworth (1983), Marcel Dekker (1984), Pergamon (1986), among others. His books are being used in over 100 countries and many of them are translated into languages such as German, Russian, Chinese, and Persian (Iranian).

He has served as general chairman of two international conferences on reliability and quality control that were held in Los Angeles and Paris in 1987. Dr. Dhillon also has served as a consultant to various organizations and has many years of experience in the industrial sector. At the University of Ottawa, he has been teaching reliability, quality, engineering management, design, and related areas for over 31 years and he has lectured in over 50 countries as well, including keynote addresses at various international scientific conferences held in North America, Europe, Asia, and Africa. In March 2004, Dr. Dhillon was a distinguished speaker at the Conference/Workshop on Surgical Errors (sponsored by the White House Health and Safety Committee and the Pentagon) held on Capitol Hill (One Constitution Avenue, Washington, D.C.).

Dr. Dhillon attended the University of Wales where he received a BS in electrical and electronic engineering and an MS in mechanical engineering. He received a PhD in industrial engineering from the University of Windsor, Ontario.

1

Introduction

1.1 Background

Each year billions of dollars are spent to develop, manufacture, operate, and maintain various types of engineering systems throughout the world. Their safety and failure due to human error have become a pressing issue because of a large number of accidental deaths and a high cost. For example, in the United States automobile accidents on highways alone cause around 42,000 deaths annually and, in 1994, the total cost of motor vehicle crashes was about $150 billion to the United States economy [1–4]. Furthermore, around 70–90% of transportation-related crashes are due to human error to a certain degree [1].

The history of safety, directly or indirectly, in regard to engineering products may be traced back to 1868 when a patent was awarded for the first barrier safeguard and to 1877 when the Massachusetts legislature passed a law requiring appropriate safeguards on hazardous machinery [5,6]. In regard to human error in engineering systems, the history goes back to the late 1950s when H. L. Williams clearly pointed out that the reliability of the involved human element must be included in the prediction of engineering systems reliability; otherwise the predicted system reliability would not depict the actual picture [7].

Over the years, a large number of publications directly or indirectly related to safety and human error in engineering systems have appeared. A list of over 500 such publications is provided in the Further Reading.

1.2 Facts and Figures

Some of the facts and figures directly or indirectly concerned with safety and human error in engineering systems are as follows:

- In 1995, work-related accidents cost the United States economy about $75 billion [8].

- Each year the United States industrial sector spends over $300 billion on plant operation and maintenance and around 80% of this figure is spent to rectify the chronic failure of machine, systems, and humans [9].
- A study of safety-related issues concerning onboard fatalities of jet fleets worldwide for the period 1982 to 1991, indicated that inspection and maintenance were clearly the second most important safety issue, with a total of 1481 onboard fatalities [10,11].
- As per Ref. [12], the annual cost of world road crashes is over $500 billion.
- Maintenance error contributes about 15% of air carrier accidents and costs the United States industrial sector over $1 billion per year [13].
- The work-related accidental deaths by cause in a typical year in the United States are poison (gas, vapor): 1.4%, water transport-related: 1.65%, poison (liquid, solid): 2.7%, air transport-related: 3%, fire-related: 3.1%, drowning: 3.2%, electric current: 3.7%, falls: 12.5%, motor vehicle-related: 37.2%, and others: 31.6% [5,14].
- In 2004, around 53% of the railway switching yard accidents (excluding highway rail crossing train accidents) in the United States were the result of human factors-related causes [15].
- In 1969, the U.S. Department of Health, Education, and Welfare special committee reported that over a period of 10 years in the United States, there were around 10,000 medical device-related injuries and 731 resulted in fatalities [16,17].
- In 2000, there were 5200 fatalities due to the occurrence of work-related accidents in the United States [6,18].
- During the period 1970–1998, about 62% of the 13 railway accidents that caused fatalities or injuries in Norway were due to human error [19].
- In 1986, the Space Shuttle Challenger exploded and all its crew members were killed [6,20].
- In 1985, a Japan Airlines Boeing 747 jet accident due to incorrect repair caused 520 fatalities [21].
- During the period 1983–1996, there were 371 major airline crashes, 29,798 general aviation crashes, and 1,735 commuter/air taxi crashes [3,23]. A study of these crashes reported that pilot error was a probable cause in 38% of major airline crashes, 85% of general aviation crashes, and 74% of commuter/air taxi crashes [3,23].
- During the period 1978–1987, there were 10 robot-related fatal accidents in Japan [24].
- A study of 6091 accident claims, over $100,000, associated with all classes of commercial ships over a period of 15 years, performed by

the U.K. P&I Club reported that 62% of the claims were attributable to human error [25–27].

- In 1986, a nuclear reactor in Chernobyl, Ukraine, exploded and directly or indirectly caused around 10,000 fatalities [6,20].
- A study reported that over 20% of all types of system failures in fossil power generation plants occur due to human errors and maintenance errors account for around 60% of the annual power loss due to human error-related problems [28].
- Human error is cited more frequently than mechanical-related problems in about 5000 truck-related fatalities that occur annually in the United States [3,29].
- Two patients died and a third patient was injured severely because of a software error in a computer-controlled therapeutic radiation machine called Therac 25 [30–32].
- In 1979, 272 people died in a DC-10 aircraft accident in Chicago because of incorrect procedures followed by maintenance workers [33].
- According to a Boeing study, in over 73% of aircraft accidents around the world, the failure of the cockpit crew has been a contributing factor [34,35].
- In 1990, a study of 126 human error-related significant events in the area of nuclear power generation revealed that around 42% of the problems were linked to modification and maintenance activities [36].
- A study of 199 human errors that occurred in Japanese nuclear power generation plants during the period 1965–1995 reported that about 50% of them were concerned with the maintenance activities [37].
- A study reported that in about 12% of major aircraft accidents, inspection and maintenance are the important factors [38,39].

1.3 Terms and Definitions

This section presents some useful terms and definitions directly or indirectly related to safety and human error in engineering systems [3,4–6,40–46].

- **Safety:** This is conservation of human life and the prevention of damage to items as per mission-specified requirements.
- **Human error:** This is the failure to perform a stated task (or the performance of a forbidden action) that could result in disruption of scheduled operations or damage to property and equipment.

- **Hazard:** This is the source of energy and the behavioral and physiological factors which, when uncontrolled, lead to harmful occurrences.
- **Human performance:** This is a measure of actions and failures under specified conditions.
- **Unsafe act:** This is an act that is not safe for an employee/individual.
- **Human factors:** This is a body of scientific facts concerning the characteristics of humans. The term includes all types of biomedical and psychosocial considerations. It also includes, but is no way restricted to, personnel selection, training principles and applications in the area of human engineering, aids for task performance, life support, and evaluation of human performance.
- **Safeguard:** This is a barrier guard, device, or procedure developed to protect humans.
- **Human performance reliability:** This is the probability that a human will satisfy all specified human functions subject to stated conditions.
- **Unsafe condition:** This is any condition, under the right set of conditions, that will lead to an accident.
- **Unsafe behavior:** This is the manner in which a person carries out actions that are considered unsafe to himself/herself or other people.
- **Accident:** This is an event that involves damage to a certain system that suddenly disrupts the current or potential system output.
- **Safety process:** This is a series of procedures followed to enable all safety requirements of an item/system to be identified and satisfied.
- **Human error consequence:** This is an undesired consequence of human failure.
- **Risk:** This is the probable occurrence rate of a hazardous condition and the degree of harm severity.
- **Continuous task:** This is a job/task that involves some kind of tracking activity (e.g., monitoring a changing condition).
- **Downtime:** This is the time during which the item/system is not in a condition to carry out its stated mission.
- **Reliability:** This is the probability that an item/system will carry out its stated function satisfactorily for the desired period when used according to the specified conditions.
- **Safety plan:** This is the implementation details of how the safety requirements of the project will be achieved.
- **Redundancy:** This is the existence of more than one means to perform a specified function.

- **Safety assessment:** This is quantitative/qualitative determination of safety.
- **Mission time:** This is that element of uptime needed to perform a stated mission profile.
- **Maintainability:** This is the probability that a failed item/system will be restored to satisfactorily operational condition.
- **Failure:** This is the inability of an item/system to carry out its specified function.

1.4 Useful Sources for Obtaining Information on Safety and Human Error in Engineering Systems

This section lists books, journals, conference proceedings, technical reports, data sources, and organizations considered useful to obtain information directly or indirectly concerned with safety and human error in engineering systems.

1.4.1 Books

- Stephans, R. A., Talso, W. W., Eds., *System Safety Analysis Handbook*, System Safety Society, Irvine, CA, 1993.
- Spellman, F. R., Whiting, N. E., *Safety Engineering: Principles and Practice*, Government Institutes, Rockville, MD, 1999.
- Dhillon, B. S., *Engineering Safety: Fundamentals, Techniques, and Applications*, World Scientific Publishing, River Edge, NJ, 2003.
- Leveson, N. G., *Safeware: System Safety and Computers*, Addison-Wesley, Reading, MA, 1995.
- Hammer, W., Price, D., *Occupational Safety Management and Engineering*, Prentice Hall, Upper Saddle River, NJ, 2001.
- Handley, W., *Industrial Safety Handbook*, McGraw Hill Book Company, London, 1969.
- Heinrich, H. W., *Industrial Accident Prevention*, 3rd ed., McGraw Hill Book Company, New York, 1950.
- Kandel, A., Avni, E., Eds., *Engineering Risk and Hazard Assessment*, CRC Press, Boca Raton, FL, 1988.
- Strauch, B., *Investigating Human Error: Incidents, Accidents, and Complex Systems*, Ashgate Publishing, Aldershot, U.K., 2002.
- Hall, S., *Railway Accidents*, Ian Allan Publishing, Shepperton, U.K., 1997.

- Dhillon, B. S., *Human Reliability, Error, and Human Factors in Engineering Maintenance*, CRC Press, Boca Raton, FL, 2009.
- Karwowski, W., Marras, W. S., *The Occupational Ergonomics Handbook*, CRC Press, Boca Raton, FL, 1999.
- Dhillon, B. S., *Human Reliability: With Human Factors*, Pergamon Press, New York, 1986.
- Sanders, M. S., McCormick, E. J., *Human Factors in Engineering and Design*, McGraw Hill Book Company, New York, 1993.
- Dhillon, B. S., *Human Reliability and Error in Transportation Systems*, Springer Inc., London, 2007.

1.4.2 Journals

- *Journal of Safety Research*
- *Safety Science*
- *Nuclear Safety*
- *Professional Safety*
- *International Journal of Reliability, Quality, and Safety Engineering*
- *Safety and Health*
- *Hazard Prevention*
- *Accident Analysis and Prevention*
- *Product Safety News*
- *Reliability Engineering and System Safety*
- *Safety Management Journal*
- *Human Factors in Aerospace and Safety*
- *Transportation Research Record*
- *Ergonomics*
- *Applied Ergonomics*
- *Human Factors*
- *International Journal of Industrial Ergonomics*
- *Human Factors and Ergonomics in Manufacturing*
- *Modern Railways*
- *International Journal of Man-Machine Studies*
- *Journal of Occupational Accidents*
- *Journal of Quality in Maintenance Engineering*
- *Risk Analysis*
- *Asia Pacific Air Safety*
- *National Safety News*

- *Safety Surveyor*
- *Air Force Safety Journal*
- *Aeronautical Journal*

1.4.3 Conference Proceedings

- Proceedings of the 48th Annual International Air Safety Seminar, 1995
- Proceedings of the International Conference on Design and Safety of Advanced Nuclear Power Plants, 1992
- Proceedings of the Human Factors and Ergonomics Society Conference, 1997
- Proceedings of the IEEE International Conference on Systems, Man, and Cybernetics, 1996
- Proceedings of the IEEE International Conference on Human Interfaces in Control Rooms, 1999
- Proceedings of the 5th Federal Aviation Administration (FAA) Meeting on Human Factors Issues in Aircraft Maintenance and Inspection, 1991
- Proceedings of the Annual Reliability and Maintainability Symposium, 2001
- Proceedings of the 15th Symposium on Human Factors in Aviation Maintenance, 2001
- Proceedings of the 7th Annual Conference on Computer Assurance, 1992
- Proceedings of the Airframe/Engine Maintenance and Repair Conference, 1998
- Proceedings of the 9th International Symposium on Aviation Psychology, 1997

1.4.4 Technical Reports

- Nuclear Power Plant Operating Experience, from the IAEA/NEA Incident Reporting System 1996–1999, Report, Organization for Economic Co-operation and Development (OECD), 2 rue Andre-Pascal, 75775 Paris Cedex 16, France, 2000.
- Seminara, J. L., Parsons, S. O., *Human Factors Review of Power Plant Maintenance*, Report No. EPRI NP-1567, Electric Power Research Institute (EPRI), Palo Alto, California, 1981.
- Moore, W. H., Bea, R. G., *Management of Human Error in Operations of Marine Systems*, Report No. HOE-93-1, 1993. Available from the

Department of Naval Architecture and Offshore Engineering, University of California, Berkeley.

- WASH-1400, *Reactor Safety Study: An Assessment of Accident Risks in U.S. Commercial Nuclear Power Plants*, U.S. Nuclear Regulatory Commission, Washington, D.C., 1975.

- Report No. DOC 9824-AN/450, *Human Factors Guidelines for Aircraft Maintenance Manual*, International Civil Aviation Organization (ICAO), Montreal, Canada, 1993.

- *Human Error in Merchant Marine Safety.* Report by the Marine Transportation Research Board, National Academy of Science, Washington, D.C., 1976.

- Report No. 5-93, *Accident Prevention Strategies, Commercial Jet Aircraft Accidents, World Wide Operations 1982–1991*, Airplane Safety Engineering Department, Boeing Commercial Airplane Group, Seattle, Washington, 1993.

- Hobbs, A., Williamson, A., *Aircraft Maintenance Safety Survey-Results*, Report, Australian Transport Safety Bureau, Canberra, Australia, 2000.

- Harvey, C. F., Jenkins, D., Sumner, R., *Driver Error*, Report No. TRRL-SR-149, Transport and Research Laboratory (TRRL), Department of Transportation, Crowthorne, U.K., 1975.

- Report No. DOT/FRA/RRS-22, *Federal Railroad Administration (FRA) Guide for Preparing Accident/Incident Reports*, FRA Office of Safety, Washington, D.C., 2003.

- Report No. PB94-917001, *A Review of Flight crew-Involved, Major Accidents of U.S. Air Carriers, 1978–1990*, National Transportation Safety Board, Washington, D.C., 1994.

1.4.5 Data Sources

- Government Industry Data Exchange Program (GIDEP), GIDEP Operations Center, U.S. Department of Navy, Corona, California.

- Gertman, D. I., Blackman, H. S., *Human Reliability and Safety Analysis Data Handbook*, John Wiley & Sons, New York, 1994.

- Boff, K. R., Lincoln, J. E., *Engineering Data Compendium: Human Perception and Performance*, Vols. 1–3, Armstrong Aerospace Medical Research Laboratory, Wright-Patterson Air Force Base, Ohio, 1988.

- Stewart, C., *The Probability of Human Error in Selected Nuclear Maintenance Tasks*, Report No. EGG-SSDC-5580, Idaho National Engineering Laboratory, Idaho Falls, Idaho, 1981.

- National Technical Information Service (NTIS), 5285 Port Royal Road, Springfield, Virginia.

- Safety Research Information Service, National Safety Council, 444 North Michigan Avenue, Chicago, Illinois.
- Nuclear Safety Information Center, Oak Ridge National Laboratory, P.O. Box Y, Oak Ridge, Tennessee.
- Computer Accident/Incident Report System, System Safety Development Center, EG&G, P.O. Box 1625, Idaho Falls, Idaho.
- Dhillon, B. S., *Human Reliability: With Human Factors*, Pergamon Press, New York, 1986. (This book lists over 20 sources for obtaining human reliability/error data.)
- Data on Equipment Used in Electric Power Generation, Equipment Reliability Information System (ERIS), Canadian Electrical Association, Montreal, Quebec, Canada.
- Dhillon, B. S., Human Error Data Banks, *Microelectronics and Reliability*, Vol. 30, 1990, pp. 963–971.
- National Maritime Safety Incident Reporting System, Maritime Administration, Washington, D.C.

1.4.6 Organizations

- World Safety Organization, P.O. Box No. 1, Lalong Laan Building, Pasay City, Metro Manila, The Philippines.
- British Safety Council, 62 Chancellors Road, London, U.K.
- U.S. Consumer Product Safety Commission, Washington, D.C.
- Board of Certified Safety Professionals, 208 Burwash Avenue, Savoy, Illinois.
- System Safety Society, 14252 Culver Drive, Suite A-261, Irvine, California.
- Transportation Research Board, 2101 Constitution Avenue NW, Washington, D.C.
- Civil Aviation Safety Authority, North Bourne Avenue and Barry Drive Intersection, Canberra, Australia.
- Airplane Safety Engineering Department, Boeing Commercial Airline Group, The Boeing Company, 7755E Marginal Way South, Seattle, Washington.
- Transportation Safety Board of Canada, 330 Spark Street, Ottawa, Ontario, Canada.
- International Civil Aviation Organization, 999 University Street, Montreal, Quebec, Canada.
- National Institute for Occupational Safety and Health (NIOSH), 200 Independence Avenue, SW Washington, D.C.

- Society for Machinery Failure Prevention Technology, 4193 Sudley Road, Haymarket, Virginia.
- Federal Railroad Administration, 4601 N. Fairfax Drive, Suite 1100, Arlington, Virginia.

1.5 Scope of the Book

Safety and human error in engineering systems have become a pressing issue because of the increasing number of accidental deaths and cost. Over the years, a large number of publications directly or indirectly related to safety and human error in engineering systems have appeared. At present, to the best of the author's knowledge, there is no specific book on the topic. This book not only attempts to cover both safety and human error in regard to engineering systems within its framework, but also provides the latest developments in the area.

Finally, the main objective of this book is to provide professionals concerned with safety and human error in engineering systems information that could be useful to improve safety and eliminate the occurrence of human error in engineering systems. This book will be useful to many individuals, including safety engineers, design engineers, system engineers, human factors engineers, and other professionals involved with engineering systems, engineering administrators and managers, psychology professionals, reliability and other engineers-at-large, researchers and instructors involved with engineering systems, and graduate and senior undergraduate students in system engineering, human factors engineering, safety, psychology, etc.

Problems

1. Define the following terms:
   a. Human error
   b. Safety
   c. Safeguard
2. Write an essay on safety and human error in engineering systems.
3. List at least four important facts and figures concerned with safety and human error in engineering systems.
4. What is the difference between the human error and unsafe behavior?
5. List at least five most important organizations for obtaining safety and human error in engineering system-related data.

6. Define the following terms:
   a. Safety process
   b. Human factors
   c. Human performance reliability
7. List at least six books considered useful for obtaining information concerned with safety and human error in engineering systems.
8. List at least five useful sources to obtain safety and human error in engineering systems-related data.
9. Define the following terms:
   a. Hazard
   b. Accident
   c. Failure
10. List at least seven journals considered most useful for obtaining safety and human error in engineering systems-related information.

References

1. Report No. 99-4, *Human-Centered Systems: The Next Challenge in Transportation*, U.S. Department of Transportation, Washington, D.C., June 1999.
2. Halll, J., Keynote Address, The American Tracking Associations Foundation Conference on Highway Accidents Litigation, September 1998. Available from the National Transportation Safety Board, Washington, D.C.
3. Dhillon, B. S., Human *Reliability and Error in Transportation Systems*, Springer, London, 2007.
4. Dhillon, B. S., *Human Reliability, Error, and Human Factors in Engineering Systems*, CRC Press, Boca Raton, FL, 2008.
5. Goetsch, D. L., *Occupational Safety and Health*, Prentice Hall, Englewood Cliffs, NJ, 1996.
6. Dhillon, B. S., *Engineering Safety: Fundamentals, Techniques, and Applications*, World Scientific Publishing, River Edge, NJ, 2003.
7. Williams, H. L., Reliability Evaluation of the Human Component in Man-Machine Systems, *Electrical Manufacturing*, April 1958, pp. 78–82.
8. Spellman, F. R., Whiting, N. E., *Safety Engineering: Principles and Practice*, Government Institutes, Rockville, MD, 1999.
9. Latino, C. J., *Hidden Treasure: Estimating Chronic Failures Can Cut Maintenance Costs up to 60%*, Report, Reliability Center, Hopewell, VA, 1999.
10. Russell, P. D., Management Strategies for Accident, *Air Asia*, Vol. 6, 1994, pp. 31–41.
11. *Human Factors in Airline Maintenance: A Study of Incident Reports*, Bureau of Air Safety Inspection, Department of Transport and Regional Development, Canberra, Australia, 1997.

12. Odero, W., *Road Traffic Injury Research in Africa: Context and Priorities*, Paper presented at the Global Forum for Health Research Conference (Forum 8), November 2004. Available from the School of Public Health, Moi University, Eldoret, Kenya.

13. Marx, D. A., *Learning from Our Mistakes: A Review of Maintenance Error Investigation and Analysis Systems* (with Recommendations to the FAA), Federal Aviation Administration (FAA), Washington, D.C., January 1998.

14. *Accident Facts: 1990–1993*, National Safety Council, Chicago, IL, 1994.

15. Reinach, S., Viale, A., Application of a Human Error Framework to Conduct Train Accident/Incident Investigations, *Accident Analysis and Prevention*, Vol. 38, 2006, pp. 396–406.

16. Banta, H. D., The Regulation of Medical Devices, *Preventive Medicine*, Vol. 19, pp. 693–699.

17. *Medical Devices*, Hearings before the Subcommittee on Public Health and Environment, U.S. Congress Interstate and Foreign Commerce, Serial No. 93-61, U.S. Government Printing Office, Washington, D.C., 1973.

18. Report on Injuries in America in 2000, National Safety Council, Chicago, IL, 2000.

19. Andersen, T., *Human Reliability and Railway Safety*, Paper presented at the Proceedings of the 16th European Safety, Reliability, and Data Association (ESREDA) Seminar on Safety and Reliability in Transport, 1999, pp. 1–12.

20. Schlager, N., *Breakdown: Deadly Technological Disasters*, Visible Ink Press, Detroit, 1995.

21. Gero, D., *Aviation Disasters*, Patrick Stephens, Sparkford, U.K., 1993.

22. *ASTB Survey of Licensed Aircraft Maintenance Engineers in Australia*, Report No. ISBN 0642274738, Australian Transport Safety Bureau (ATSB), Department of Transport and Regional Services, Canberra, Australia, 2001.

23. Fewer Airline Crashes Linked to "Pilot Error," Inclement Weather Still Major Factor, *Science Daily*, January 9, 2001.

24. Nagamachi, M., *Ten Fatal Accidents Due to Robots in Japan, Ergonomics of Hybrid Automated Systems*, edited by W. Karwowski et al., Elsevier, Amsterdam, 1988, pp. 391–396.

25. Just Waiting to Happen—The Work of the U.K. P&I Club, *The International Maritime Human Element Bulletin*, No. 1, October 2003, pp. 3–4. Published by the Nautical Institute, London.

26. DVD Spotlights Human Error in Shipping Accidents, *Asia Maritime Digest*, January/February 2004, pp. 41–42.

27. Boniface, D. E., Bea, R. G., Assessing the Risk of and Countermeasures for Human and Organizational Error, *SNAME Transactions*, Vol. 104, 1996, pp. 157–177.

28. Daniels, R. W., *The Formula for Improved Plant Maintainability Must Include Human Factors*, Paper presented at the proceedings of the IEEE Conference on Human Factors and Nuclear Safety, 1985, pp. 242–244.

29. Truck Safety Snag: Handling Human Error, *The Detroit News*, Detroit, July 17, 2000.

30. Schneider, P., Hines, M. L. A., *Classification of Medical Software*, Paper presented at the proceedings of the IEEE Symposium on Applied Computing, 1990, pp. 20–27.

31. Gowen, L. D., Yap, M. Y., *Traditional Software Development's Effects on Safety*, Paper presented at the proceedings of the 6th Annual IEEE Symposium on Computer-Based Medical Systems, 1993, pp. 58–63.
32. Joyce, E., Software Bugs: A Matter of Life and Liability, *Datamation*, Vol. 33, No. 10, 1987, pp. 88–92.
33. Christensen, J. M., Howard, J. M, *Field Experience in Maintenance, in Human Detection and Diagnosis of System Failures*, edited by J. Rasmussen and W. B. Rouse, Plenum Press, New York, 1981, pp. 111–133.
34. Majos, K., Communication and Operational Failures in the Cockpit, *Human Factors and Aerospace Safety*, Vol. 1, No. 4, 2001, pp. 323–340.
35. Report No. 1-96, *Statistical Summary of Commercial Jet Accidents: Worldwide Operations: 1959–1996*, Boeing Commercial Airplane Group, Seattle, WA, 1996.
36. Reason, J., Human Factors in Nuclear Power Generation: A System's Perspective, *Nuclear Europe Worldscan*, Vol. 17, No. 5–6, 1997, pp. 35–36.
37. Hasegawa, T., Kemeda, A., *Analysis and Evaluation of Human Error Events in Nuclear Power Plants*, Presented at the meeting of the IAEA's CRP on Collection and Classification of Human Reliability Data for Use in Probabilistic Safety Assessments, May 1998. Available from the Institute of Human Factors, Nuclear Power Engineering Corporation, 3-17-1 Toranomon, Minato-Ku, Tokyo.
38. Marx, D. A., Graeber, R. C., *Human Error in Maintenance*, in *Aviation Psychology in Practice*, edited by N. Johnston, N. McDonald, and R. Fuller, Ashgate Publishing, London, 1994, pp. 87–104.
39. Gray, N., *Maintenance Error Management in the ADF*, Touchdown (Royal Australian Navy), December 2004, pp. 1–4. Available online at http://www.navy.gov.au/publications/touchdown/dec.04/maintrr.html
40. *Dictionary of Terms Used in the Safety Profession*, 3rd ed., American Society of Safety Engineers, Des Plaines, IL, 1988.
41. Meulen, M. V. D., *Definitions for Hardware and Software Safety Engineers*, Springer-Verlag, London, 2000.
42. IEEE-STD-1228, *Standard for Software Safety Plans*, Institute of Electrical and Electronic Engineers (IEEE), New York, 1994.
43. Dhillon, B. S., *Human Reliability: With Human Factors*, Pergamon Press, New York, 1986.
44. MIL-STD-721B, *Definitions of Effectiveness Terms for Reliability, Maintainability, Human Factors, and Safety*, Department of Defense, Washington, D.C., August 1966. Available from the Naval Publications and Forms Center, 801 Tabor Ave., Philadelphia, PA.
45. MIL-STD-1908, *Definitions of Human Factors Terms*, Department of Defense, Washington, D.C.
46. McKenna, T., Oliverson, R., *Glossary of Reliability and Maintenance Terms*, Gulf Publishing Company, Houston, TX, 1997.

2

Basic Mathematical Concepts

2.1 Introduction

Just like in the development of any other area of engineering, mathematics also has played an important role in the development of safety and human factors/error fields. The history of mathematics may be traced back to the development of our currently used number symbols often called the "Hindu–Arabic numeral system" [1]. The first evidence of the use of these numerals or symbols is found on stone columns erected by the Scythian Emperor of India named Asoka, in 250 BCE [1].

The earliest reference to the probability concept may be traced back to the writing of a gambler's manual by Girolamo Cardano (1501–1576) [1,2]. However, Pierre Fermat (1601–1665) and Blaise Pascal (1623–1662) were the first two persons who solved correctly and independently the problem of dividing the winnings in a game of chance. Boolean algebra, which plays an important role in modern probability theory, is named after the mathematician George Boole (1815–1864), who published, in 1847, a pamphlet entitled "The Mathematical Analysis of Logic: Being an Essay towards a Calculus of Deductive Reasoning" [1,3].

Needless to say, a more detailed history of mathematics and probability is available in Refs. [1,2]. This chapter presents basic mathematical concepts considered useful to understanding subsequent chapters of this book.

2.2 Range, Median, Arithmetic Mean, and Mean Deviation

A given set of safety or human error-related data are useful only if it is analyzed effectively. More specifically, there are certain characteristics of the data that are quite helpful in describing the nature of a given dataset, thus making better associated decisions. Thus, this section presents four statistical measures considered useful to study safety and human error in engineering systems-related data [4,5].

2.2.1 Range

This is a useful measure of dispersion or variation. The range of values in a dataset is the difference between the largest and the smallest values in the set.

Example 2.1

Assume that the following set of numbers represent human errors in engineering systems occurring over a 12-month period in an organization:

$$5, 20, 16, 4, 25, 15, 7, 10, 22, 18, 9, \text{ and } 26$$

Find the range of the given dataset.

By examining the given dataset, we conclude that the largest and the smallest values are 26 and 4, respectively. Thus, the range of the dataset values is

$$R = \text{Largest value} - \text{Smallest value}$$

$$= 26 - 4$$

$$= 22$$

where
  R = the range.

It means that the range of the given dataset values is 22.

2.2.2 Median

The median of a set of data values arranged in an array (i.e., in order of magnitude) is the middle value or the average of the very two middle values.

Example 2.2

Arrange Example 2.1 dataset values in an array (i.e., in order of magnitude) and then find the dataset median.

Thus, an array of the dataset values is as follows:

$$4, 5, 7, 9, 10, 15, 16, 18, 20, 22, 25, \text{ and } 26$$

The very two middle values of the above array of values are 15 and 16. Thus, the average of these two values is 15.5. It means the median of the dataset is 15.5.

2.2.3 Arithmetic Mean

This is defined by:

$$m = \frac{\sum\limits_{i=1}^{n} m_i}{n} \tag{2.1}$$

where
  m = the mean value (i.e., arithmetic mean)
  m_i = the data value i, for i = 1, 2, 3, ..., n
  n = the total number of data values

It is to be noted that often arithmetic mean is simply referred to as mean.

Example 2.3

Assume that the quality control department of an engineering systems manufacturing company inspected six identical systems and found 2, 4, 6, 8, 3, and 10 defects in each system due to human error. Calculate the mean or average number of defects per system (i.e., arithmetic mean) due to human error.
  By inserting the specified data values into equation (2.1), we obtain:

$$m = \frac{2+4+6+8+3+10}{6}$$

$$= 5.5$$

Thus, the mean number of defects per system due to human error is 5.5. In other words, the arithmetic mean of the given dataset is 5.5.

2.2.4 Mean Deviation

This is a measure of dispersion and its value indicates the degree to which a given set of data tend to spread about a mean value. Mean deviation is defined by:

$$MD = \frac{\sum\limits_{i=1}^{n} |m_i - m|}{n} \tag{2.2}$$

where
  n = the total number of data values
  MD = the mean deviation

m = the mean value of the given dataset
m_i = the data value i; for i = 1, 2, 3, ..., n
$|m_i - m|$ = the absolute value of the deviation of m_i for m

Example 2.4

Calculate the mean deviation of the dataset given in Example 2.3.

The calculated mean value of the dataset from Example 2.3 is $m = 5.5$ defects per system.

Using the above calculated value and the given data values in equation (2.2), we obtain:

$$MD = \frac{|2-5.5|+|4-5.5|+|6-5.5|+|8-5.5|+|3-5.5|+|10-5.5|}{6}$$

$$= \frac{[3.5+1.5+0.5+2.5+2.5+4.5]}{6}$$

$$= 2.5$$

Thus, the mean deviation of the dataset given in Example 2.3 is 2.5.

2.3 Boolean Algebra Laws

Boolean algebra is used to a degree in engineering systems-related safety and human error studies and is named after Boole, its founder. Some of the Boolean algebra laws are presented below [3,6].

$$X.Y = Y.X \qquad (2.3)$$

where
  Y = an arbitrary set or event
  X = an arbitrary set or event

Dot (.) denotes the intersection of sets. Sometime equation (2.3) is written without the dot (e.g., XY), but it still conveys the same meaning.

$$X + Y = Y + X \qquad (2.4)$$

where
  + = the union of sets or events

$$XX = X \qquad (2.5)$$

$$X + X = X \qquad (2.6)$$

$$(XY)Z = X(YZ) \qquad (2.7)$$

where
  Z = an arbitrary set or event

$$(X + Y) + Z = X + (Y + Z) \qquad (2.8)$$

$$X + XY = X \qquad (2.9)$$

$$X(X + Y) = X \qquad (2.10)$$

$$X(Y + Z) = XY + YZ \qquad (2.11)$$

$$(X + Y)(Y + Z) = X + YZ \qquad (2.12)$$

2.4 Probability Definition and Properties

The probability is defined as follows [7]:

$$P(Y) = \lim_{m \to \infty}\left(\frac{M}{m}\right) \qquad (2.13)$$

where
  $P(Y)$ = the probability of occurrence of event Y
  M = the number of times event Y occurs in the m repeated experiments

Some of the basic properties of probability are as follows [7,8]:

- The probability of occurrence of event, say X, is

$$0 \le P(X) \le 1 \qquad (2.14)$$

- Probability of sample space S is

$$P(S) = 1 \qquad (2.15)$$

- Probability of the negation of the sample space S is

$$P(\bar{S}) = 0 \qquad (2.16)$$

- The probability of occurrence and nonoccurrence of an event, say X, is always:

$$P(X) + P(\bar{X}) = 1 \qquad (2.17)$$

where
  $P(X)$ = the probability of occurrence of event X
  $P(\bar{X})$ = the probability of nonoccurrence of event X

- The probability of an intersection of n independent events is

$$P(X_1 X_2 X_3 \ldots X_n) = P(X_1)(P(X_2)P(X_3) \ldots P(X_n) \qquad (2.18)$$

where
  $P(X_i)$ = the probability of occurrence of event X_i, for i = 1, 2, 3, ... n.

- The probability of the union of n independent events is given by:

$$P(X_1 + X_2 + \ldots + X_n) = 1 - \prod_{i=1}^{n} (1 - P(X_i)) \qquad (2.19)$$

For n = 2, Equation (2.19) reduces to

$$P(X_1 + X_2) = P(X_1) + P(X_2) - P(X_1)P(X_2) \qquad (2.20)$$

- The probability of the union of n mutually exclusive events is

$$P(X_1 + X_2 + \ldots + X_n) = \sum_{i=1}^{n} P(X_i) \qquad (2.21)$$

2.5 Basic Probability Distribution-Related Definitions

This section presents a number of probability distribution-related definitions considered useful to conduct various types of safety and human error
studies concerned with engineering systems.

2.5.1 Probability Density Function

For a continuous random variable, the probability density function is defined by:

$$f(t) = \frac{dF(t)}{dt} \tag{2.22}$$

where
 $F(t) =$ the cumulative distribution function
 $t =$ time (i.e., a continuous random variable)
 $f(t) =$ the probability density function

2.5.2 Cumulative Distribution Function

For a continuous random variable, the cumulative distribution function is defined by:

$$F(t) = \int_{-\infty}^{t} f(x)\,dx \tag{2.23}$$

For $t = \infty$, equation (2.23) becomes:

$$F(\infty) = \int_{-\infty}^{\infty} f(x)\,dx \tag{2.24}$$

$$= 1$$

It means that the total area under the probability density curve is equal to unity. Usually, in safety and human error work, equation (2.23) is simply written as:

$$F(t) = \int_{0}^{t} f(x)\,dx \tag{2.25}$$

2.5.3 Expected Value

The expected value of a continuous random variable is defined by:

$$E(t) = \int_{-\infty}^{\infty} tf(t)\,dt \tag{2.26}$$

where
 $E(t) =$ the expected value (i.e., mean value) of the continuous random variable t

2.6 Probability Distributions

This section presents a number of probability or statistical distributions considered useful to perform various types of safety and human error studies concerned with engineering systems.

2.6.1 Exponential Distribution

This is one of the simplest continuous random variable distributions widely used in the industrial sector, particularly in reliability and safety studies [9]. The probability density function of the distribution is defined by:

$$f(t) = \lambda e^{-\lambda t}, t \geq 0, \lambda > 0 \tag{2.27}$$

where
  $f(t)$ = the probability density function
  λ = the distribution parameter
  t = time (i.e., a continuous random variable)

Inserting equation (2.27) into equation (2.25) yields the following expression for the cumulative distribution function:

$$F(t) = 1 - e^{-\lambda t} \tag{2.28}$$

With the aid of equation (2.26) and equation (2.27), we obtain the following expression for the distribution expected value (i.e., mean value):

$$E(t) = \frac{1}{\lambda} \tag{2.29}$$

2.6.2 Rayleigh Distribution

This continuous random variable distribution is named after its founder, John Rayleigh (1842–1919) [1]. The distribution is often used in the theory of sound and time to time in reliability and safety studies as well. The probability density function of the distribution is defined by:

$$f(t) = \left(\frac{2}{\alpha^2} \right) t e^{-\left(\frac{t}{\alpha} \right)^2}, \ t \geq 0, \ \alpha \rangle 0 \tag{2.30}$$

where
  α = the distribution parameter

Substituting equation (2.30) into equation (2.25), we obtain the following expression for the cumulative distribution function:

$$F(t) = 1 - e^{-\left(\frac{t}{\alpha}\right)^2} \tag{2.31}$$

With the aid of equation (2.26) and equation (2.30), we get the following expression for the distribution expected value:

$$E(t) = \alpha\Gamma\left(\frac{3}{2}\right) \tag{2.32}$$

where
$\Gamma(.)$ = the gamma function and is defined by:

$$\Gamma(z) = \int_0^\infty t^{z-1} e^{-t} \, dt, \text{ for } z > 0 \tag{2.33}$$

2.6.3 Weibull Distribution

This is another continuous random variable distribution and it was developed by Wallodi Weibull, a Swedish mechanical engineering professor, in the early 1950s [10]. The probability density function of the distribution is defined by:

$$f(t) = \frac{at^{a-1}}{\alpha^a} e^{-\left(\frac{t}{\alpha}\right)^a} \quad t \geq 0, \alpha\rangle 0, a\rangle 0 \tag{2.34}$$

where
α and a = the scale and shape parameters, respectively

Using equation (2.25) and equation (2.34), we obtain the following cumulative distribution function:

$$F(t) = 1 - e^{-\left(\frac{t}{\alpha}\right)^a} \tag{2.35}$$

Substituting equation (2.34) into equation (2.26), we obtain the following expression for the Weibull distribution expected value:

$$E(t) = \alpha\Gamma\left(1 + \frac{1}{a}\right) \tag{2.36}$$

It is to be noted that for $a = 1$ and $a = 2$, the exponential and Rayleigh distributions are the special cases of this distribution, respectively.

2.6.4 Normal Distribution

This is one of the most widely known continuous random variable distributions and sometimes it is called the Gaussian distribution, after a German mathematician, Carl Friedrich Gauss (1777–1855). The probability density function of the distribution is defined by:

$$f(t) = \frac{1}{\sigma\sqrt{2\pi}} \exp\left[-\frac{(t-\mu)^2}{2\sigma^2}\right], \ \infty\langle t\langle+\infty \tag{2.37}$$

where
μ and σ = the distribution parameters (i.e., mean and standard deviation, respectively)

Substituting equation (2.37) into equation (2.23), we get the following equation for cumulative distribution function:

$$F(t) = \frac{1}{\sigma\sqrt{2\pi}} \int_{-\infty}^{t} \exp\left[-\frac{(x-\mu)^2}{2\sigma^2}\right] dx \tag{2.38}$$

Using equation (2.26) and equation (2.37), we get the following expression for the distribution expected value:

$$E(t) = \frac{1}{\sigma\sqrt{2\pi}} \int_{-\infty}^{\infty} t\exp\left[-\frac{(t-\mu)^2}{2\sigma^2}\right] dt \tag{2.39}$$

$$= \mu$$

2.7 Laplace Transform Definition, Common Laplace Transforms, Final Value Theorem Laplace Transform, and Laplace Transforms' Application in Solving First-Order Differential Equations

This section presents various aspects of Laplace transforms considered useful to perform safety and human error studies concerned with engineering systems.

2.7.1 Laplace Transform Definition

The Laplace transform of the function f(t) is defined by:

$$f(s) = \int_{0}^{\infty} f(t)e^{-st} \, dt \tag{2.40}$$

where
  $f(s)$ = the Laplace transform of function $f(t)$
  t and s = the time and Laplace transform variables, respectively

Example 2.5

Obtain the Laplace transform of the following function:

$$f(t) = 1 \tag{2.41}$$

By substituting equation (2.41) into equation (2.40), we obtain:

$$f(s) = \int_0^\infty 1 . e^{-st} \, dt$$

$$= \frac{e^{-st}}{-s} \Big|_0^\infty \tag{2.42}$$

$$= \frac{1}{s}$$

Example 2.6

Obtain the Laplace transform of the following function:

$$f(t) = e^{-\theta t} \tag{2.43}$$

where
  θ = a constant

Using equation (2.40) and equation (2.43), we get:

$$f(s) = \int_0^\infty e^{-\theta t} e^{-st} \, dt$$

$$= \frac{e^{-(s+\theta)t}}{-(s+\theta)} \Big|_0^\infty \tag{2.44}$$

$$= \frac{1}{s+\theta}$$

2.7.2 Laplace Transforms of Common Functions

Laplace transforms of some commonly occurring functions in safety and human error studies concerned with engineering systems are presented in Table 2.1 [11].

TABLE 2.1

Laplace transforms of some commonly occurring
functions in safety and human error studies

No.	$f(t)$	$f(s)$
1	$e^{-\theta t}$	$\dfrac{1}{(s+\theta)}$
2	t^m, for m = 0, 1, 2, 3, ...	$\dfrac{m!}{s^{m+1}}$
3	C, a constant	$\dfrac{C}{s}$
4	$te^{-\theta t}$	$\dfrac{1}{(s+\theta)^2}$
5	$\dfrac{df(t)}{dt}$	$sf(s)-f(0)$
6	$\lambda_1 f_1(t) + \lambda_2 f_2(t)$	$\lambda_1 f_1(s) + \lambda_2 f_2(s)$
7	$tf(t)$	$-\dfrac{df(s)}{ds}$

2.7.3 Final Value Theorem Laplace Transform

If the following limits exist, then the final-value theorem may be stated as:

$$\lim_{t \to \infty} f(t) = \lim_{s \to 0} \left[sf(s) \right] \tag{2.45}$$

2.7.4 Laplace Transforms' Application in Solving First-Order Differential Equations

Usually, Laplace transforms are used to find solutions to linear first-order differential equations in safety and human error studies concerned with engineering systems. The following example demonstrates the finding of solutions to a set of linear first-order differential equations describing a human error problem in engineering systems:

Example 2.7

Assume that an engineering system can be in any of the three states: operating normally, failed due to a human error, or failed due to a hardware failure. The following three linear first-order differential equations describe the system:

$$\frac{dP_0(t)}{dt} + (\lambda_{he} + \lambda_{hf})P_0(t) = 0 \tag{2.46}$$

$$\frac{dP_1(t)}{dt} - \lambda_{he} P_0(t) = 0 \tag{2.47}$$

$$\frac{dP_2(t)}{dt} - \lambda_{hf} P_0(t) = 0 \tag{2.48}$$

where

$P_j(t)$ = the probability that the system is in state j at time t, for $j = 0$ (operating normally), $j = 1$ (failed due to a human error), and $j = 2$ (failed due to a hardware failure)

λ_{he} = the system constant human error rate

λ_{hf} = the system constant hardware failure rate

At time $t = 0$, $P_0(0) = 1$, $P_1(0) = 0$, and $P_2(0) = 0$.

Solve differential equation (2.46) to equation (2.48) by using Laplace transforms.

Using Table 2.1, differential equation (2.46) through equation (2.48), and the specified initial conditions, we obtain:

$$sP_0(s) - 1 + (\lambda_{he} + \lambda_{hf})P_0(s) = 0 \tag{2.49}$$

$$sP_1(s) - \lambda_{he} P_0(s) = 0 \tag{2.50}$$

$$sP_2(s) - \lambda_{hf} P_0(s) = 0 \tag{2.51}$$

Solving equation (2.49) through equation (2.51), we get:

$$P_0(s) = \frac{1}{(s + \lambda_{he} + \lambda_{hf})} \tag{2.52}$$

$$P_1(s) = \frac{\lambda_{he}}{s(s + \lambda_{he} + \lambda_{hf})} \tag{2.53}$$

$$P_2(s) = \frac{\lambda_{hf}}{s(s + \lambda_{he} + \lambda_{hf})} \tag{2.54}$$

The inverse Laplace transforms of equation (2.52) to equation (2.54) are as follows:

$$P_0(t) = e^{-(\lambda_{he} + \lambda_{hf})t} \tag{2.55}$$

$$P_1(t) = \frac{\lambda_{he}}{\lambda_{he} + \lambda_{hf}} \left[1 - e^{-(\lambda_{he} + \lambda_{hf})t} \right] \qquad (2.56)$$

$$P_2(t) = \frac{\lambda_{hf}}{\lambda_{he} + \lambda_{hf}} \left[1 - e^{-(\lambda_{he} + \lambda_{hf})t} \right] \qquad (2.57)$$

Thus, equation (2.55) to equation (2.57) are the solutions to differential equation (2.46) to equation (2.48).

Problems

1. Define the following two items:
   a. Range
   b. Median
2. Assume that the quality control department of an engineering systems manufacturer inspected eight identical systems and found 4, 7, 5, 3, 1, 10, 12, and 6 defects in each system due to human error. Calculate the average or mean number of defects per system due to human error.
3. Calculate the mean deviation of the dataset given in the above problem (i.e., problem no. 2).
4. Prove the Boolean algebra expression (2.12).
5. Define the following items:
   a. Cumulative distribution function
   b. Expected value
6. What are the special case distributions of the Weibull distribution?
7. Prove equation (2.31).
8. Define the following:
   a. Laplace transform
   b. Probability
9. Prove that the sum of equation (2.55) through equation (2.57) is equal to unity.
10. Write down the probability density functions for the following two probability distributions:
    a. Normal distribution
    b. Exponential distribution

References

1. Eves, H., *An Introduction to the History of Mathematics*, Holt, Rinehart and Winston, New York, 1976.
2. Owen, D. B., Ed., *On the History of Statistics and Probability*, Marcel Dekker, New York, 1976.
3. Lipschutz, S., *Set Theory*, McGraw Hill Book Company, New York, 1964.
4. Spiegel, M. R., *Probability and Statistics*, McGraw Hill Book Company, New York, 1975.
5. Speigel, M. R., *Statistics*, McGraw Hill Book Company, New York, 1961.
6. *Fault Tree Handbook*, Report No. NUREG-0492, U.S. Nuclear Regulatory Commission, Washington, D.C., 1981.
7. Mann, N. R., Schafer, R. E., Singpurwalla, N. D., *Methods for Statistical Analysis of Reliability and Life Data*, John Wiley & Sons, New York, 1974.
8. Lipschutz, S., *Probability*, McGraw Hill Book Company, New York, 1965.
9. Davis, D. J., An Analysis of Some Failure Data, *J. Amer. Stat. Assoc.*, June 1952, pp. 113–150.
10. Weibull, W., A Statistical Distribution Function of Wide Applicability, *J. Appl. Mech.*, Vol. 18, March 1951, pp. 293–297.
11. Oberhettinger, F., Badic, L., *Tables of Laplace Transforms*, Springer-Verlag, New York, 1973.

3

Safety and Human Factors and Error Basics

3.1 Introduction

Today safety has become an important field and its history in the modern times may be traced back to 1868 when a patent was awarded for the first barrier safeguard in the United States [1]. In 1893, the United States Congress passed the Railway Safety Act and, in 1912, the cooperative Safety Congress met for the first time in Milwaukee, Wisconsin [1,2]. Nowadays, the field of safety has developed into many areas including medical equipment safety, software safety, transportation systems safety, and robot safety.

Over the years many new developments have taken place in the areas of human factors and error, and in many parts of the world both human factors and error have become recognizable disciplines in the industrial sector. The history of human factors may be traced back to 1898 when Frederick W. Taylor performed a number of studies for determining the most effective designs for shovels [3]. Similarly, the history of human error in regard to engineering systems may be traced back to 1958 when H. L. Williams clearly pointed out that the reliability of the human element must be considered in the overall prediction of the system reliability; otherwise the predicted system reliability would not depict the actual picture [4]. Nowadays, the field of human error has developed into many areas including transportation systems, healthcare systems, and power systems.

This chapter presents various introductory safety and human factors and error concepts considered useful to understand subsequent chapters of this book, which have been taken from published literature.

3.2 Safety and Engineers and Product Hazard Classifications

The problem of safety with engineering systems may be traced back to railroads. For example, a prominent English legislator was killed in a railroad accident the very day Stephenson's first railroad line was dedicated [2]. The following year, the boiler of the first locomotive built in the United States exploded and caused one fatality and badly injured several fuel servers [2,5]. Needless to say, modern engineering products have become highly sophisticated and complex and their safety is a challenging issue because of competition and other factors when engineers are pressured to complete new designs rapidly and at minimum cost. Past experiences over the years clearly indicate that this, in turn, normally led to more design deficiencies, errors, and causes of accidents.

Design deficiencies can cause or contribute to accidents and they may occur because a designer/design [2,6]:

- Overlooked to eliminate or reduce the occurrence of human errors.
- Relies on product users to avoid the occurrence of accidents.
- Is confusing, wrong, or unfinished.
- Creates an unsafe characteristic of a product/item.
- Incorporates weak warning mechanisms instead of providing a safe design to eradicate potential hazards.
- Overlooked to prescribe effective operational procedures in situations where hazards might exist.
- Violates usual tendencies/capabilities of users.
- Overlooked to warn properly of potential hazards.
- Overlooked to provide adequate protection in a user's personal protective equipment.
- Overlooked to foresee unexpected applications of an item/product or its all potential consequences.
- Places an unreasonable degree of stress on potential operators.
- Creates an arrangement of operating controls and other devices that increases reaction time in emergency conditions or is conducive to the occurrence of errors.
- Does not properly consider or determine the action, error, omission, or failure consequences.

There are many product hazards. They may be grouped under the following six classifications [6,7]:

- **Human factors hazards:** These are concerned with poor design with respect to people. More specifically, to their physical strength, visual angle, weight, height, intelligence, visual acuity education, computational ability, length of reach, etc.
- **Energy hazards:** These are of two types: kinetic energy hazards and potential energy hazards. The kinetic energy hazards pertain to parts such as flywheels, fan blades, and loom shuttles because of their motion. Any object that interferes with the motion of such parts can

experience substantial damage. The potential energy hazards pertain to parts such as springs, electronic capacitors, counterbalancing weights, and compressed gas receivers that store energy. During the equipment servicing process such hazards are very important because stored energy can result in serious injury when released suddenly.

- **Kinematic hazards:** These hazards are associated with scenarios where parts come together while still moving and lead to possible crushing, cutting, or pinching of any object caught between them.
- **Electrical hazards:** These hazards have two principal components: shock hazard and electrocution hazard. The major electrical hazard to property/product stems from electrical faults, often called *short circuits*.
- **Misuse- and abuse-related hazards:** These hazards are concerned with the usage of products by people. Past experiences indicate that misuse of products can lead to serious injuries. Product abuse also can result in hazardous situations or injuries and some of the causes for the abuse are lack of necessary maintenance and poor operating practices.
- **Environmental hazards:** These are of two types: internal and external. The internal hazards are concerned with the changes in the surrounding environment that result in internally damaged product. This type of hazard can be eradicated or minimized by considering factors such as vibrations, electromagnetic radiation, extremes of temperatures, illumination level, atmospheric contaminants, and ambient noise levels during the design process. The external hazards are the hazards posed by the product during its life span and include maintenance hazards, services-life operation hazards, and disposal hazards.

3.3 Common Mechanical Injuries and Common Causes of Product Liability Exposure

In the day-to-day work environment in the industrial sector, people interact with various types of engineering equipment to perform tasks, such as drilling, shaping, cutting, stamping, punching, abrading, and chipping. There are various types of injuries that can occur during the performance of such tasks; some of the common ones are as follows [1]:

- **Shearing-related injuries:** These injuries are concerned with shearing processes. In manufacturing, power-driven shears are commonly used to perform tasks such as severing metal, plastic, elastomers, and paper. In the past, time-to-time during the use of such machines tragedies, such as amputation of hands/fingers, have occurred.

- **Breaking-related injuries:** These injuries are usually concerned with machines used to deform engineering materials. Frequently, a break in a bone is referred to as a fracture. In turn, fracture is categorized under many classifications including complete, oblique, simple, transverse, compound, comminuted, and incomplete.
- **Puncturing-related injuries:** These injuries occur when an object penetrates straight into a person's body and pulls straight out. In the industrial sector, normally, such injuries pertain to punching machines because they contain sharp tools.
- **Crushing-related injuries:** These injuries occur where a person's body part is caught between two hard surfaces moving progressively together and crushing any item/object that comes between them.
- **Cutting- and tearing-related injuries:** These injuries occur when a person's body part comes in contact with a sharp edge. The severity of a cut or tear depends upon the degree of damage to skin, veins, muscles, arteries, etc.
- **Straining- and spraining-related injuries:** For the occurrence of these injuries (e.g., straining of muscles or spraining of ligaments), in the industrial environment, there are numerous opportunities associated with the use of machines or other tasks.

Past experiences over the years indicate that about 60% of the liability cases involved failure to provide proper danger warning labels on manufactured items/products. Nonetheless, some of the common causes of product liability exposure include [8]:

- Faulty product design
- Faulty manufacturing
- Poorly written warnings
- Inadequate testing of product prototypes
- Poorly written instructions
- Inadequate research during product development

3.4 Safety Management Principles and Product Safety Organization Tasks

There are many principles of safety management. The main ones are as follows [9,10]:

- The safety system should be tailored to fit the company culture.

- Safety should be managed just like managing any other function in an organization. More clearly, management should direct safety by setting attainable safety goals, and by organizing, planning, and controlling to successfully attain the goals.
- There is no single method for achieving safety in an organization. But, for a safety system to be effective, it must meet certain criteria: be flexible, involve workers participation, have the top level management visibly showing its support, etc.
- In building a good safety system, the three major subsystems that must be considered with care are the physical, the behavioral, and the managerial.
- The key to successful line safety performance is management procedures that clearly factor in accountability.
- The causes leading to unsafe behavior can be controlled, identified, and classified.
- There are certain sets of circumstances that can be predicted to lead to severe injuries: nonproductive activities, certain construction situations, high energy sources, and unusual, nonroutine tasks.
- The three important symptoms that indicate that something is not right in the management system are an accident, an unsafe act, and an unsafe condition.
- The main function of safety is to find and define the operational errors that result in accidents.
- Under most circumstances, unsafe behavior is normal behavior because it is the result of normal human beings reacting to their surrounding environment. Therefore, it is clearly the responsibility of management to carry out necessary changes to the environment that leads to the unsafe behavior.

A product safety organization performs many tasks. The main ones include [8,11]

- Review all safety-related field reports and customer complaints.
- Provide assistance to designers in choosing alternative means to eradicate or control hazards or other safety-related problems in preliminary designs.
- Review nongovernmental and governmental product safety-related requirements.
- Review proposed product maintenance and operation documents in regard to safety.
- Prepare the product safety program and directives.

- Develop a system by which the safety program can be monitored effectively.
- Determine if items, such as emergency equipment, protective equipment, or monitoring and warning devices, are really required for the product.
- Review the product to determine if all the potential hazards have been properly controlled or eradicated.
- Review warning labels that are to be placed on the product in regard to compatibility to warnings in the instruction manuals, adequacy, and meeting all legal requirements.
- Take part in reviewing accident claims or recall actions by government bodies and recommend appropriate remedial actions for justifiable claims or recalls.
- Prepare safety criteria on the basis of applicable voluntary and governmental standards for use by vendor, subcontractor, and company design professionals.
- Review all types of mishaps and hazards in current similar product to avoid their repetition in the new products.
- Review product test reports for determining deficiencies or trends in regard to safety.

3.5 Accident Causation Theories

There are many accident causation theories including the human factors theory, the domino theory, the systems theory, the epidemiological theory, the combination theory, and the accident/incident theory [1,8]. The first two of these theories are described below.

3.5.1 The Human Factors Theory

This theory is based on the assumption that accidents occur due to a chain of events caused by human error. There are three main factors that cause human errors [1,12]:

1. Overload
2. Inappropriate response/incompatibility
3. Inappropriate activities

The factor "overload" is concerned with an imbalance between a person's capacity at any time and the load he/she is carrying in a given state.

The capacity of a person is the product of a number of factors including fatigue, stress, natural ability, state of mind, degree of training, and physical condition. The load carried by a person is made up of tasks for which he/she has responsibility, along with additional burdens due to the following factors [8,12]:

- **Environmental factors:** Two examples of these factors are noise and distractions.
- **Situational factors:** Two examples of these factors are unclear instructions and level of risk.
- **Internal factors:** Three examples of these factors are worry, emotional stress, and personal problems.

The factor "inappropriate response/incompatibility" is another main human error causal factor and some examples of inappropriate response by a person include [8,12]:

- A person totally removed a safeguard from a machine/equipment to improve output.
- A person detected a hazardous condition, but took no appropriate corrective measure.
- A person totally disregarded the recommended safety-related procedures.

Finally, the factor "inappropriate activities" is concerned with inappropriate activities performed by a person due to human error. For example, a person misjudged the degree of risk involved in a specified task and then carried out the task on that misjudgment.

Additional information on the human factors theory is available in Ref. [12].

3.5.2 The Domino Theory

This theory is operationalized in 10 statements known as the "Axioms of Industrial Safety." These axioms were developed by H. W. Heinrich (American industrial safety pioneer) and include [1,8,13]:

Axiom I: Supervisors are the key persons in the prevention of industrial accidents.

Axiom II: Most accidents are due to the unsafe acts of people.

Axiom III: Management should assume safety responsibility with full vigor because it is in the best position to achieve final results.

Axiom IV: The reasons why people commit unsafe acts can be quite useful in selecting necessary corrective actions.

Axiom V: The severity of an injury is largely fortuitous and the specific accident that caused it is normally preventable.

Axiom VI: An unsafe condition or an unsafe act by a person does not always immediately result in an injury/accident.

Axiom VII: The occurrence of injuries results from a completed sequence of events or factors, the final one of which is the accident itself.

Axiom VIII: An accident can occur only when a person commits an unsafe act and/or there is some physical or mechanical hazard.

Axiom IX: The most effective accident prevention approaches are quite analogous with the quality and productivity methods.

Axiom X: There are direct and indirect costs of an accident. Some examples of the direct cost are medical costs, compensation, and liability claims.

Furthermore, as per Heinrich, there are five factors in the sequence of events leading up to an accident [1,8,13]:

Factor I: Ancestry and social environment. In this case, it is assumed that negative character traits, such as stubbornness, avariciousness, and recklessness that might lead humans to behave unsafely, can be inherited through ancestry or acquired as a result of the social surroundings.

Factor II: Fault of person. In this case, it is assumed that negative character traits (i.e., whether acquired or inherited), such as recklessness, violent temper, nervousness, and ignorance of safety practices, constitute proximate reasons to commit unsafe acts or for the presence of physical or mechanical hazards.

Factor III: Unsafe act/mechanical or physical hazard. In this case, it is assumed that unsafe acts, such as removing safeguards, starting machinery without warning, and standing under suspended loads, committed by humans, and physical or mechanical hazards, such as inadequate light, unguarded gears, absence of rail guards, and unguarded point of operation, are the direct causes for the occurrence of accidents.

Factor IV: Accident. In this case, it is assumed that events, such as falls of people and striking of people by flying objects, are examples of accidents that lead to injury.

Factor V: Injury. In this case, it is assumed that the injuries directly caused by accidents include lacerations and fractures.

All in all, the following two items are the central points of the Heinrich theory [8,12]:

1. Injuries are the result of the action of all preceding factors.
2. The eradication of the central factor, i.e., unsafe act/hazardous condition, definitely negates the action of preceding factors and, in turn, prevents accidents and injuries.

Additional information on the Domino theory is available in Ref. [12].

3.6 Human Factors Objectives and Man–Machine Comparisons

There are many human factors objectives. They may be grouped under four classifications as shown in Figure 3.1 [14,15]. These classifications are fundamental operational objectives, objectives affecting reliability and maintainability, objectives affecting operators and users, and miscellaneous objectives.

The fundamental operational objectives are basically concerned with improving safety, reducing human errors, and improving system performance. The objectives affecting reliability and maintainability are concerned with items such as improving reliability, reducing the manpower need, improving maintainability, and reducing training requirements. The objective affecting operators and users are items such as increasing

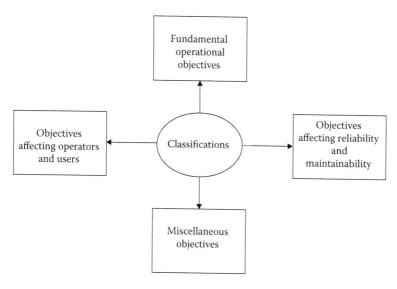

FIGURE 3.1
Classifications of human factors objectives.

user acceptance and ease of use, increasing aesthetic appearance, improving the work environment, and reducing boredom, physical stress, fatigue, and monotony.

Finally, the miscellaneous objectives are concerned with items such as increasing production economy and reducing equipment and time losses.

Time to time to make certain engineering decisions knowledge about comparisons between humans and machines become quite essential. In this regard, some of the important comparisons between humans and machines (in parentheses) are: [15,16]

- Humans are highly flexible in regard to task performance (machines are relatively inflexible).
- Humans are subjected to social environments of all types (machines are independent of social environments of all kinds).
- Humans possess excellent memory (machines are extremely expensive for that same capability).
- Humans are limited in channel capacity (machines have unlimited channel capacities).
- Humans are unsuitable to perform tasks such as data coding, transformation, or amplification (machines are extremely suitable to perform such tasks).
- Performance efficiency of humans is affected by anxiety (machines are independent of this problem).
- Humans are highly capable of making inductive decisions in novel conditions (machines possess little or no induction capabilities).
- Humans are subjected to departure from following an optimum strategy (machines always execute the designed strategy).
- Humans have relatively easy maintenance (maintenance problems of machines become serious with the increase in complexity).
- Humans possess high tolerance for factors such as uncertainty, ambiguity, and vagueness (machines are very limited in tolerance with respect to such factors).
- Humans possess extremely limited short-term memory for factual matters (machines can have unlimited short-term memory, but its affordability is a limiting factor).
- Humans are prone to stress because of interpersonal or other problems (machines are independent of such problems).
- Humans are prone to factors such as disorientation, motion sickness, and coriolis effects (machines are independent of such effects).
- Humans are subjected to degradation of performance because of fatigue and boredom (machines are not subjected to such factors, but their performance can degrade due to wear or lack of calibration).

- Humans are not good monitors of events that occur infrequently (machines have options to be designed to reliably detect events that do not occur frequently).

Additional information on man–machine comparisons is available in Ref. [16].

3.7 Typical Human Behaviors and Human Sensory Capacities

Over the years professionals working in the area have identified many typical human behaviors. Some of these behaviors include [15,17]:

- Humans generally regard manufactured product as being safe.
- Often humans tend to hurry.
- Humans usually expect to turn on the electrical power, the switches have to move upward, or to the right, etc.
- Humans get easily confused with unfamiliar things.
- Humans expect that faucets and valve handles will rotate counter-clockwise to increase the flow of gas, liquid, or steam.
- Humans have become accustomed to specific color meanings.
- Humans frequently use their hands first to test or explore the unknown.
- Attention of humans is drawn to factors such as flashing lights, loud noises, bright lights, and bright and vivid colors.

Humans possess many useful sensors: sight, hearing, touch, taste, and smell. More specifically, humans can sense pressure, temperature, vibration, rotation, position, linear motion, and acceleration (shock). A clear understanding of their (humans) sensory capacities can be very useful in reducing human errors in engineering systems. Thus, some of the human sensory-related capacities are discussed below [15,18].

3.7.1 Sight

Sight is stimulated by the electromagnetic radiation of certain wavelengths, frequently referred to as the visible segment of the electromagnetic spectrum. The various areas of the spectrum, as seen by the eyes, appear to vary in brightness. For example, during the day time, the human eyes are very sensitive to greenish-yellow light with a wavelength of around 5500 Angstrom units [18]. Also, the eyes see differently from different angles.

Moreover, usually the eyes perceive all colors when they are looking straight ahead; however, the color perception decreases with the increase in the viewing angle.

3.7.2 Touch

This is closely related to humans' ability to interpret visual and auditory stimuli. The sensory cues received by muscles and the skin can be used to send messages to the brain, thus relieving the human eyes and ears of the workload, to a certain degree. Additional information on touch is available in Ref. [16].

3.7.3 Noise

Noise may simply be described as sounds that lack coherence. The reaction of humans to the problem of noise extends beyond the auditory system. It can result in feelings such as irritability, boredom, or fatigue. Excessive noise can lead to various problems including loss in hearing if exposed for long periods, adverse effects on tasks requiring a high degree of muscular coordination or intense concentration, and reduction in a worker's efficiency.

3.7.4 Vibration

Past experiences over the years clearly indicate that the presence of vibration could be detrimental to the performance of mental and physical tasks by humans. There are many vibration parameters: frequency, velocity, amplitude, jolt, and acceleration. More specifically, low frequency and large amplitude vibrations contribute to fatigue, headaches, motion sickness, eye strain, and deterioration in ability to read and interpret instruments [18]. Furthermore, high frequency and low amplitude vibrations can be quite fatiguing.

3.8 Useful Human Factors Guidelines and Mathematical Human Factors-Related Formulas

Over the years, professionals working in the area of human factors have developed many human factors guidelines considered useful for application in designing engineering systems. Some of these guidelines are presented in Table 3.1 [15–17].

In the published literature, there are various types of mathematical formulas to estimate human factors-related information. Some of these formulas considered useful for studying, directly or indirectly, human error in engineering systems are presented below.

TABLE 3.1

Useful Human Factors Guidelines

No.	Guideline
1	Review system objectives in regard to human factors
2	Fabricate a hardware prototype (if possible) and evaluate this under applicable environments
3	Review final production drawings with respect to human factors
4	Conduct field tests of the system design prior to approving it for delivery to customers/users
5	Use mock-ups for "testing" the effectiveness of all user-hardware interface designs
6	Develop a human factors checklist for use in design and production phases
7	Acquire applicable human factors-related design reference documents
8	Perform experiments when cited reference guides fail to provide necessary information for design-related decisions
9	Use the services of human factors specialists as considered appropriate

3.8.1 Formula I: Rest Period

This formula is concerned with estimating the length of the rest period required for humans performing various types of tasks. The length of the required rest is expressed by [15,19,20]:

$$R = \frac{WT(AE - SC)}{(AE - \theta)} \tag{3.1}$$

where
   R = the required rest expressed in minutes
   WT = the working time expressed in minutes
   θ = the approximate resting level expressed in kilocalories per minute (usually, the value of θ is taken as 1.5)
   AE = the average energy expenditure/cost expressed in kilocalories per minute work
   SC = the kilocalories per minute adopted as standard

Example 3.1

Assume that an engineering systems maintenance worker is performing a certain maintenance task for 100 minutes and his/her average energy expenditure is 4 kilocalories per minute. Calculate the length of the required rest period for the worker if SC = 3 kilocalories per minute.

By substituting the given data values into equation (3.1), we get:

$$R = \frac{(100)(4-3)}{(4-1.5)}$$

$$= 40 \text{ minutes}$$

Thus, the length of the required rest period is 40 minutes.

3.8.2 Formula II: Character Height

This formula is concerned with estimating the character height at the viewing distance of 28 inches, as usually the instrument panels are located at a viewing distance of 28 inches for the comfortable performance and control of adjustment-oriented tasks. Thus, the character height is expressed by [21,22].

$$CH = \frac{(SCH)(RVD)}{28} \tag{3.2}$$

where
RVD = the required viewing distance expressed in inches
CH = the character height at the required viewing distance (RVD), expressed in inches
SCH = the standard character height at a viewing distance of 28 inches

Example 3.2

Assume that an engineering systems operator has to read a meter from a distance of 60 inches and the standard character height at a viewing of 28 inches at low luminance is 0.40 inch. Estimate the height of numerals for the specified viewing distance.

By substituting the specified data values into equation (3.2), we obtain:

$$CH = \frac{(0.40)(60)}{28}$$

$$= 0.86 \text{ inch}$$

Thus, the height of numerals for the specified viewing distance is 0.86 inch.

3.8.3 Formula III: Glare Constant

This formula is concerned with estimating the value of the glare constant as human errors can occur, in performing engineering systems-related operation and maintenance tasks, due to glare. The glare constant is expressed by [15,19]:

$$\beta = \frac{(\theta^{1.6})(\lambda^{0.8})}{(GBL)\mu^2} \qquad (3.3)$$

where
β = the glare constant
θ = the source luminance
μ = the angle between the direction of the glare source and the viewing direction
λ = the solid angle subtended at the eye by the source
GBL = the general background luminance

Additional information on the glare constant is available in Ref. [19].

3.8.4 Formula IV: Inspector Performance

This formula is concerned with estimating inspector performance in regard to inspection-oriented tasks. The inspector performance is expressed by [15,23]:

$$\alpha = \frac{RT}{NP - NE} \qquad (3.4)$$

where
α = the inspector performance expressed in minutes per correct inspection
RT = the reaction time in minutes
NP = the number of patterns inspected
NE = the number of inspector errors

Additional information on this formula is available in Ref. [23].

3.9 Human Error Occurrence Examples and Studies, and Reasons

Over the years human error has been responsible for many accidents and equipment/system failures. In fact, many studies have clearly reported a significant proportion of equipment and other failures due to human error. The findings of some of those studies include:

- A total of 401 human errors occurred in U.S. commercial light-water nuclear reactors during the period from June 1, 1973 to June 30, 1975 [24].
- A study of 135 vessel failures occurring during the period from 1926 to 1988 reported that 24.5% of the failures were directly the result of human error [25].
- A study reported that over 90% of the documented air traffic control system errors were due to human operators [26].
- A study of 23,000 defects in the production of nuclear parts reported that about 82% of the defects were the result of human error [27].
- Up to 90% of accidents both generally and in medical devices are due to human error [28,29].
- Over 50% of all technical medical equipment problems are the result of operator errors [30].

In general, there could be many reasons for the occurrence of human errors. Some of the important ones are poor equipment design, poor work layout, complex tasks, poor motivation of involved personnel, inadequate work tools, poor training or skill of concerned personnel, inadequate or poorly written equipment operating and maintenance procedures, and poor job environment (i.e., poor lighting, high noise level, crowded work space, high/low temperature, etc.) [31,32].

3.10 Human Error Types

Human errors may be broken into seven types as shown in Figure 3.2 [31–33]. These types are operator errors, design errors, assembly errors, inspection errors, installation errors, handling errors, and maintenance errors.

Operator errors are the result of operator mistakes, and the conditions that lead to operator errors include operator carelessness, poor personnel selection and training, complex tasks, poor environments, and lack of proper procedures. Design errors are the result of poor design. The causes of these errors include failure to ensure the man–machine interaction effectiveness, assigning inappropriate functions to humans, and failure to implement human needs in the design. An example of design errors is the placement of displays and controls so far apart that operators are unable to use them effectively.

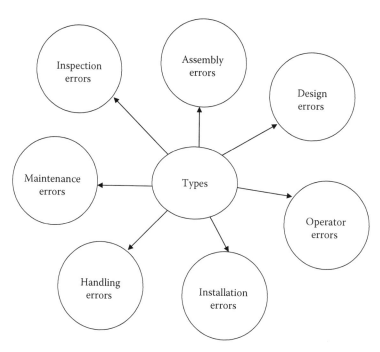

FIGURE 3.2
Types of human errors.

Assembly errors occur during the product assembly process due to humans. These errors may occur due to causes, such as poor blue prints, excessive noise level, poorly designed work layout, poor communication of related information, inadequate illumination, and excessive temperature in the work area. Inspection errors occur because of less than 100% inspectors' accuracy. An average inspection effectiveness is around 85% [34]. One typical example of an inspection error is rejecting and accepting in-tolerance and out-of-tolerance parts, respectively.

Installation errors occur due to various reasons including failing to install equipment according to the manufacturer's specification or using the incorrect installation blueprints or instructions. Handling errors basically occur because of poor transportation or storage facilities. More specifically, such facilities are not as stated by the equipment manufacturers. Finally, maintenance errors occur in the field environment due to oversights by the maintenance personnel. As the equipment becomes old, the likelihood of the occurrence of such errors may increase because of the increase in the frequency of maintenance. Three examples of the maintenance errors are repairing the failed equipment incorrectly, applying the incorrect grease at appropriate points of equipment, and calibrating equipment incorrectly.

3.11 General Stress Factors and Occupational Stressors

Over the years, professionals working in the area of human factors have pointed out that there are many general factors that increase stress on humans, in turn leading to a significant deterioration in their reliability. Some of these general factors are poor health, serious financial difficulties, poor chances for promotion, possibility of redundancy at work, excessive demands from superiors at work, lacking the proper expertise to perform the ongoing job, working under extremely tight time pressures, having to work with individuals with unpredictable temperaments, and experiencing difficulties with spouse or children or both [15,32].

There are many occupational stressors. They may be grouped under four classifications as shown in Figure 3.3 [35]. Workload-related stressors are concerned with work overload or work underload. In the case of work overload, the job requirements exceed an individual's ability to satisfy them effectively. In contrast, in the case of work underload, the present duties being performed by the person fail to provide sufficient stimulation. Three examples of work underload are the lack of opportunity to apply acquired skills and expertise of the person, the lack of any intellectual input, and task repetitiveness. Occupational frustration-related stressors are concerned with the problems related to occupational frustration. The problems include the ambiguity of one's role, inadequate career development guidance, and the lack of proper communication.

Occupational change-related stressors are concerned with factors that disrupt physiological, cognitive, and behavioral patterns of functioning of the

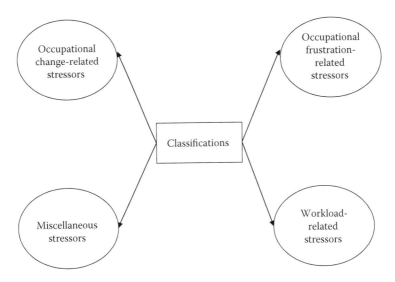

FIGURE 3.3
Classifications of occupational stressors.

person. Finally, miscellaneous stressors include all those stressors that are not included into the above three classifications. Some examples of these stressors are too little or too much lighting, too much noise, and poor interpersonal relationships.

Additional information on occupational stressors is available in Ref. [35].

Problems

1. Describe the following three types of product hazards:
   a. Energy hazards
   b. Human factors hazards
   c. Environmental hazards
2. What are the commonly occurring mechanical injuries?
3. List at least six common causes of product liability exposure.
4. What are the main principles of safety management?
5. What are the main tasks performed by a product safety organization?
6. Describe the Domino theory.
7. Make at least 10 comparisons between humans and machines.
8. What are the typical human behaviors?
9. What are the main reasons for the occurrence of human errors?
10. Describe the following five types of human errors:
    a. Design errors
    b. Maintenance errors
    c. Operator errors
    d. Inspection errors
    e. Assembly errors

References

1. Goetsch, D. L., *Occupational Safety and Health*, Prentice Hall, Englewood Cliffs, NJ, 1996.
2. Hammer, W., Price, D., *Occupational Safety Management and Engineering*, Prentice Hall, Upper Saddle River, NJ, 2001.
3. Chapanis, A., *Man-Machine Engineering*, Wadsworth Publishing Company, Belmont, CA, 1965.

4. Williams, H. L., Reliability Evaluation of the Human Component in Man-Machine Systems, *Electrical Manufacturing*, April 1958, pp. 78–82.
5. Hunter, T. A., Operator Safety, *Engineering*, May 1974, pp. 358–363.
6. Dhillon, B. S., *Reliability, Quality, and Safety for Engineers*, CRC Press, Boca Raton, FL, 2005.
7. Hunter, T. A., *Engineering Design for Safety*, McGraw-Hill Book Company, New York, 1992.
8. Dhillon, B. S., *Engineering Safety: Fundamentals, Techniques, and Applications*, World Scientific Publishing, River Edge, NJ, 2003.
9. Petersen, D., *Safety Management*, American Society of Safety Engineers, Des Plaines, IL, 1998.
10. Petersen, D., *Techniques of Safety Management*, McGraw-Hill Book Company, New York, 1971.
11. Hammer, W., *Product Safety Management and Engineering*, Prentice Hall, Inc., Englewood Cliffs, NJ, 1980.
12. Heinrich, H. W., Petersen, D., Roos, N., *Industrial Accident Prevention*, McGraw-Hill Book Company, New York, 1980.
13. Heinrich, H. W., *Industrial Accident Prevention*, 4th ed., McGraw-Hill Book Company, New York, 1959.
14. Chapanis, A., *Human Factors in Systems Engineering*, John Wiley & Sons, New York, 1996.
15. Dhillon, B. S., *Human Reliability, Error, and Human Factors in Engineering Maintenance*, CRC Press, Boca Raton, FL, 2009.
16. Dhillon, B. S., *Advanced Design Concepts for Engineers*, Technomic Publishing Company, Lancaster, PA, 1998.
17. Woodson, W. E., *Human Factors Design Handbook*, McGraw-Hill Book Company, New York, 1981.
18. AMCP-706-134, *Engineering Design Handbook: Maintainability Guide for Design*, prepared by the U.S. Army Material Command, Alexandria, VA, 1972.
19. Oborne, D. J., *Ergonomics at Work*, John Wiley & Sons, New York, 1982.
20. Murrell, K. F. H., *Human Performance in Industry*, Reinhold Publishing Company, New York, 1965.
21. Huchingson, R. D., *New Horizons for Human Factors in Design*, McGraw-Hill Book Company, New York, 1981.
22. McCormick, E. J., Sanders, M. S., *Human Factors in Engineering and Design*, McGraw-Hill Book Company, New York, 1982.
23. Drury, C. G., Fox, J. G., Eds., *Human Reliability in Quality Control*, John Wiley & Sons, New York, 1975.
24. Joos, D. W., Sabril, Z. A., Husseiny, A. A., Analysis of Gross Error Rates in Operation of Commercial Nuclear Power Stations, *Nuclear Engineering Design*, Vol. 52, 1979, pp. 265–300.
25. *Organizational Management and Human Factors in Quantitative Risk Assessment*, Report No. 33/1992 (Report 1), British Health and Safety Executive (HSE), London, 1992.
26. Kenney, G. C., Spahn, M. J., Amato, R. A., *The Human Element in Air Traffic Control: Observations and Analysis of Performance of Controllers and Supervisors in Providing Air Traffic Control Separation Services*, METREK Div., MITRE Corp., Report No. MTR-7655, Bedford, MA, December 1977.

27. Rook, L. W., *Reduction of Human Error in Industrial Production*, Report No. SCTM 93-63 (14), Sandia Laboratories, Albuquerque, NM, June 1962.

28. Nobel, J. L., Medical Devices Failures and Adverse Effects, *Pediatric-Emergency Care*, Vol. 7, 1991, pp. 120–123.

29. Askern, W. B., Regulinski, T. L., Quantifying Human Performance for Reliability Analysis of Systems, *Human Factors*, Vol. 11, 1969, pp. 393–396.

30. Dhillon, B. S., Reliability Technology in Health Care Systems, paper presented at the Proceedings of IASTED, International Symposium of Computers and Advanced Technological Medicine, *Health Care, Bioengineering*, 1990, pp. 84–87.

31. Meister, D., The Problem of Human-Initiated Failures, paper presented at the Proceedings of the Eighth National Symposium on Reliability and Quality Control, 1962, pp. 234–239.

32. Dhillon, B. S., *Human Reliability: With Human Factors*, Pergamon Press, New York, 1986.

33. Cooper, J. L., Human Initiated Failures and Man-Function Reporting, *IRE Trans. Human Factors*, Vol. 10, 1961, pp. 104–109.

34. McCornack, R. L., *Inspector Accuracy: A Study of the Literature*, Report No. SCTM 53-61 (14), Sandia Corporation, Albuquerque, NM, 1961.

35. Beech, H. R., Burns, L. E., Sheffield, B. F., *A Behavioral Approach to the Management of Stress*, John Wiley & Sons, New York, 1982.

4

Methods for Performing Safety and Human Error Analysis in Engineering Systems

4.1 Introduction

Over the years, a large number of publications in the fields of safety, human factors, and reliability have appeared in the form of journal articles, conference proceedings articles, books, and technical reports [1–4]. Many of these publications report the development of various types of methods and approaches to perform safety, human factors, and reliability analyses. Some of these methods and approaches can be used to perform safety and human error analysis in engineering systems. The others are more confined to a specific field (i.e., safety, human factors, or reliability).

Two important examples of these methods and approaches that can be used to perform safety and human error analysis in engineering systems are failure modes and effect analysis (FMEA) and fault tree analysis (FTA). FMEA was developed in the early 1950s by the U.S. Department of Defense to analyze engineering systems from the reliability aspect. Nowadays, FMEA is being used in many diverse areas including safety, human factors, management, and healthcare [2,4–7]. FTA was developed in the early 1960s at the Bell Telephone Laboratories to perform safety and reliability analysis of the Minuteman Launch Control System [7–9]. Nowadays, FTA is being used to analyze various types of problems in a wide range of areas including engineering, management, and healthcare [1–4,7–9].

This chapter presents a number of methods and approaches considered useful in performing safety and human error analysis in engineering systems, extracted from the published literature in the areas of safety, human factors, and reliability.

4.2 Interface Safety Analysis (ISA)

ISA is concerned with determining the incompatibilities between assemblies and subsystems of a product/item that could result in accidents. The analysis establishes that totally distinct parts/units can be integrated into a quite viable system and the normal operation of an individual part or unit will not deteriorate the performance or damage another part/unit or the entire product/system. Although ISA considers various relationships, they can be grouped basically under three categories, as shown in Figure 4.1 [10].

The functional relationships are concerned with multiple items or units. For example, in a situation where a unit's outputs constitute the inputs to a downstream unit, an error in outputs and inputs may cause damage to the downstream unit and, in turn, become a safety hazard. The condition of outputs could be:

- Excessive outputs
- Zero outputs
- Degraded outputs
- Erratic outputs
- Unprogrammed outputs

The physical relationships are connected to the physical aspects of items or units. For example, two items or units might be well designed and manufactured and operate quite well individually, but they may fail to fit together effectively due to dimensional differences or they may present

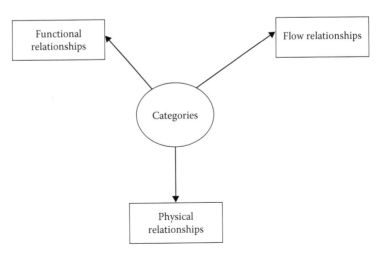

FIGURE 4.1
Categories of relationships considered by ISA.

other problems that may lead to safety-related issues. Three typical examples of the other problems include [2,10]:

- Impossible to tighten, join, or mate parts/components properly.
- Impossible or restricted access to or egress from equipment.
- A very little clearance between units; thus the units may be damaged during replacement or removal process.

Finally, the flow relationships may involve two or more units or items. For example, the flow between two units or items may entail lubricating oil, steam, water, air, electrical energy, or fuel. In addition, the flow also could be unconfined, such as heat radiation from one body to another. Normally, the common problems associated with many products are the proper flow of energy and fluids from one item or unit to another through confined passages or spaces, consequently leading to direct or indirect safety-related issues/problems. Nonetheless, the flow associated problem causes include complete or partial interconnection failure and faulty connections between units. In the case of fluid, factors, such as listed below, must be considered seriously from the safety aspect [2,10].

- Toxicity
- Flammability
- Lubricity
- Loss of pressure
- Contamination
- Corrosiveness
- Odor

4.3 Technic of Operations Review

This method was developed in the early 1970s by D.A. Weaver of the American Society of Safety Engineers (ASSE) and seeks to identify systemic causes rather than assigning blame in regard to safety [2,11,12]. Technic of operations review (TOR) allows employees and management to work jointly in performing analysis of workplace related incidents, accidents, and failures, and it may be described simply as a hands-on analytical tool for identifying the root system causes of an operation failure.

The method uses a worksheet containing simple terms, basically, requiring "yes/no" decisions. An incident occurring at a certain point in time and location involving certain individuals activates TOR. Furthermore, it may be added that TOR demands systematic evaluation of the actual circumstances surrounding the incident as it is not a hypothetical process.

The following eight steps are associated with TOR [2,11,12]:

Step 1: Form the TOR team by choosing its members from all concerned areas.

Step 2: Hold a roundtable meeting to depart common knowledge to all team members.

Step 3: Highlight one key systemic factor that played an instrumental role in the occurrence of accident/incident. This factor serves as a starting point for further investigation and it must be based on the team consensus.

Step 4: Use team consensus in responding to a sequence of "yes/no" options.

Step 5: Evaluate the highlighted factors by ensuring the clear existence of team consensus in regard to the evaluation of each factor.

Step 6: Prioritize all the contributory factors by starting with the most serious factor.

Step 7: Develop appropriate corrective/preventive strategies in regard to each and every contributory factor.

Step 8: Implement the strategies.

All in all, the main strength of the TOR is the involvement of line personnel in the analysis and its main weakness is an after-the-fact process.

4.4 Root Cause Analysis

This method was developed by the U.S. Department of Energy to investigate industrial incidents, and it may simply be described as a systematic investigation approach that uses information collected during an assessment of an accident to determine the underlying factors for deficiencies that caused the accident [13,14].

The 10 general steps involved in performing root cause analysis (RCA) include [15,16]:

Step 1: Educate all individuals involved in RCA.

Step 2: Inform appropriate staff members when a sentinel event is reported.

Step 3: Form an RCA team composed of appropriate individuals.

Step 4: Prepare for and hold the first team meeting.

Step 5: Determine the event sequence.

Step 6: Separate and highlight each event sequence that may have been a contributory factor in the sentinel event occurrence.

Step 7: Brainstorm about the factors surrounding the selected events that may have been contributory to the sentinel event occurrence.

Step 8: Affinitize with the brainstorming session results.

Step 9: Develop the action plan.

Step 10: Distribute the action plan and the RCA document to all individuals concerned.

Over the years, RCA has been applied in many areas. Some of the advantages and disadvantages observed with the application of this method are as follows [16,17]:

Advantages

- It is a well structured and process-focused approach.
- The systematic application of the method can uncover common root causes that link a disparate collection of accidents.
- It is an effective tool for identifying and addressing systems and organizational issues.

Disadvantages

- It is a labor intensive and time-consuming method.
- It is impossible to determine exactly if the root cause established by the analysis is really the actual cause for the occurrence of the accident.
- In essence, RCA is basically an uncontrolled case study.
- It is quite possible to be tainted by hindsight bias.

A list of the available RCA software packages in the market is given in Ref. [16].

4.5 Hazards and Operability Analysis

This method was developed for application in the chemical industry and is considered a powerful tool for identifying safety-related problems prior to availability of complete data concerning an item [18,19]. Three fundamental objectives of hazards and operability (HAZOP) analysis include [11,20–21]:

- To produce a complete description of facility/process.
- To review each facility/process part to determine how deviations from the design intentions can occur.
- To decide whether the deviations can result in operating hazards/problems.

The seven steps involved in performing HAZOP include [11,19,22]:

Step 1: Select the process/system to be analyzed.

Step 2: Form the team of appropriate individuals (experts).

Step 3: Explain the HAZOP process to all team members.

Step 4: Establish appropriate goals and time schedules.

Step 5: Conduct brainstorming sessions.

Step 6: Conduct analysis.

Step 7: Document the study.

Finally, it is to be noted that HAZOP has basically the same weaknesses as failure modes and effect analysis (FMEA) discussed below in Section 4.7. For example, both of these methods predict problems that are connected to system/process failures, but overlook factoring human error into the equation. This is a very important weakness because human error is quite often a factor in the occurrence of accidents.

4.6 Preliminary Hazard Analysis

This is relatively an unstructured method because of the unavailability of definitive data or information, such as functional flow diagrams and drawings. Thus, it is basically used during the concept design phase. Nonetheless, over the years, preliminary hazard analysis (PHA) has proved to be a quite useful approach to take in the early steps of identifying and eliminating hazards when the desirable data are unavailable. The findings of the PHA are extremely helpful to serve as a useful guide in potential detailed analysis.

PHA requires the formation of an ad hoc team made up of members familiar with items such as equipment, material, substance, and/or process. The members of the team are asked to review hazard occurrences in the area of their expertise and, as a team, play the devil's advocate. Additional information on PHA is available in Ref. [23].

4.7 Failure Modes and Effect Analysis (FMEA)

This is a widely used method during the design process to analyze engineering systems from their reliability aspect. It simply may be described as an effective approach to analyze each potential mode in the system to examine the effects of such failure modes on the system [24]. FMEA also can be applied to conducting engineering systems safety analysis.

The history of FMEA may be traced back to the early years of the 1950s with the development of flight control systems, when the U.S. Navy's Bureau of Aeronautics, in order to develop a procedure for reliability control over the detail design effort, developed a requirement known as Failure Analysis [25]. Subsequently, the term *Failure Analysis* was changed to FMEA and in the 1970s, the U.S. Department of Defense directed its effort to developing a military standard entitled "Procedures for Performing a Failure Mode, Effects, and Criticality Analysis" [26]. Basically, failure mode, effects, and criticality analysis (FMECA) is an extended version of FMEA. More specifically, when FMEA is extended to group each potential failure effect in regard to its severity (this includes documenting critical and catastrophic failures), the method is known as FMECA [2,27].

The seven main steps followed to perform FMEA are shown in Figure 4.2 [9]. There are many factors that must be explored before the implementation of FMEA. Some of these factors include [6,28]:

- Examination of each and every conceivable failure mode by the involved professionals
- Measuring FMEA cost/benefits
- Obtaining engineer's approval and support
- Making decisions based on the risk priority number (RPN).

Over the years professionals working in the area of reliability analysis have developed a number of guidelines/facts concerning FMEA. Some of the guidelines/facts include [2,6]:

- FMEA is not the tool for selecting the optimum design concept.
- FMEA has certain limitations.
- Avoid developing the majority of the FMEA in a meeting.
- FMEA is not designed to supersede the engineer's work.
- RPN could be misleading.

Nonetheless, there are many advantages of performing FMEA. Some of the main ones are shown in Figure 4.3 [1,6].

Additional information on FMEA is available in Ref. [1].

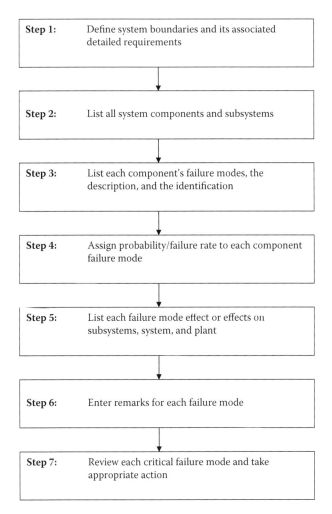

FIGURE 4.2
Main steps followed to perform FMEA.

4.8 Probability Tree Method

This method is used to conduct task analysis by diagrammatically representing critical human actions and other events associated with the system under consideration. Frequently, the method is utilized to conduct task analysis in the technique for human error rate prediction (THERP) [3,29].

In this approach, diagrammatic task analysis is represented by the branches of the probability tree. The branching limbs of the tree represent

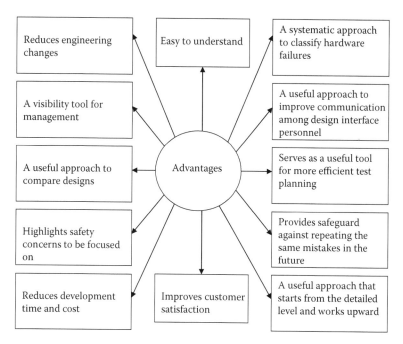

FIGURE 4.3
FMEA advantages.

each event's outcome (i.e., success or failure) and each branch is assigned an occurrence probability.

Some of the advantages of the probability tree method are [9,30]

- Includes a useful visibility tool
- Decreases the probability of errors due to computation because of computational simplification
- Incorporates, with some modifications, factors such as emotional stress, interaction stress, and interaction effects.

The following two examples demonstrate the application of the probability tree method.

4.8.1 Examples of Probability Tree Method

Example 4.1

Assume that an engineering technician performs three independent safety-related tasks: a, b, and c. Each of these three tasks can be performed either correctly or incorrectly and task a is performed before task b, and task b before task c.

Develop a probability tree and obtain an expression for the probability of not successfully accomplishing the overall mission by the engineering technician.

In this case, the engineering technician first performs task a correctly or incorrectly and then proceeds to carry out task b. Task b also can be carried out correctly or incorrectly by the engineering technician. After task b, the technician proceeds to carry out task c. This task can be carried out correctly or incorrectly as well by the engineering technician. This entire scenario is shown in Figure 4.4.

The meanings of the symbols used in Figure 4.4 are defined as: a = the task a is performed correctly, $\bar{a}$ = the task a is performed incorrectly, b = the task b is performed correctly, $\bar{b}$ = the task b is performed incorrectly, c = the task c is performed correctly, and $\bar{c}$ = the task c is performed incorrectly.

By examining the Figure 4.4 diagram, it can be concluded that there are a total of seven possibilities (i.e., $ab\bar{c}$, $a\bar{b}c$, $a\bar{b}\bar{c}$, $\bar{a}bc$, $\bar{a}b\bar{c}$, $\bar{a}\bar{b}c$, and $\bar{a}\bar{b}\bar{c}$) for not successfully accomplishing the overall mission by the engineering technician. Thus, the probability of not successfully accomplishing the overall mission by the engineering technician is given by:

$$P_{nt} = P(ab\bar{c}) + P(a\bar{b}c) + P(a\bar{b}\bar{c}) + P(\bar{a}bc) + P(\bar{a}b\bar{c}) + P(\bar{a}\bar{b}c) + P(\bar{a}\bar{b}\bar{c})$$

(4.1)

$$= P_a P_b P_{\bar{c}} + P_a P_{\bar{b}} P_c + P_a P_{\bar{b}} P_{\bar{c}} + P_{\bar{a}} P_b P_c + P_{\bar{a}} P_b P_{\bar{c}} + P_{\bar{a}} P_{\bar{b}} P_c + P_{\bar{a}} P_{\bar{b}} P_{\bar{c}}$$

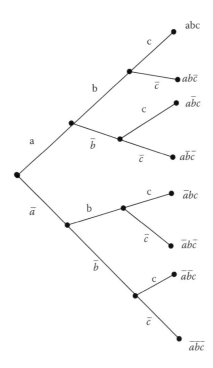

FIGURE 4.4
Probability tree diagram for the engineering technician performing tasks a, b, and c.

where

P_{nt} = the probability of not successfully accomplishing the overall mission by the engineering technician

P_a = the probability of performing task a correctly by the engineering technician

$P_{\bar{a}}$ = the probability of performing task a incorrectly by the engineering technician

P_b = the probability of performing task b correctly by the engineering technician

$P_{\bar{b}}$ = the probability of performing task b incorrectly by the engineering technician

P_c = the probability of performing task c correctly by the engineering technician

$P_{\bar{c}}$ = the probability of performing task c incorrectly by the engineering technician.

Thus, equation (4.1) is the expression for the probability of not successfully accomplishing the overall mission by the engineering technician.

Example 4.2

If the last task (i.e., task c) in Example 4.1 is eliminated, develop a probability tree and probability expression for successfully accomplishing the overall mission by the engineering technician. In addition, calculate the probability of not successfully accomplishing the overall mission by the engineering technician if the probabilities of performing tasks a and b correctly are 0.9 and 0.8, respectively.

As there are only two tasks performed by the engineering technician, the probability tree of Figure 4.4 reduces to the one shown in Figure 4.5.

With the aid of this diagram, the probability of successfully accomplishing the overall mission by the engineering technician is

$$P_{st} = P(ab) = P_a P_b \qquad (4.2)$$

where

P_{st} = the probability of successfully accomplishing the overall mission by the engineering technician

Similarly, with the aid of the Figure 4.5 diagram, the probability of not successfully accomplishing the overall mission by the engineering technician is

$$P_{nt} = P(a\bar{b}) + P(\bar{a}b) + P(\bar{a}\bar{b})$$
$$= P_a P_{\bar{b}} + P_{\bar{a}} P_b + P_{\bar{a}} P_{\bar{b}} \qquad (4.3)$$

where

P_{nt} = the probability of not successfully accomplishing the overall mission by the engineering technician

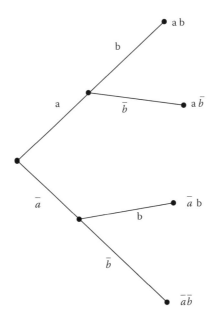

FIGURE 4.5
Probability tree diagram for the engineering technician performing tasks a and b only.

Because $P_{\bar{a}} = 1 - P_a$ and $P_{\bar{b}} = 1 - P_b$, equation (4.3) reduces to:

$$P_{nt} = 1 - P_a P_b \qquad (4.4)$$

By substituting the given data values into equation (4.4), we get:

$$P_{nt} = 1 - (0.9)(0.8)$$
$$= 0.28$$

Thus, the probability of not successfully accomplishing the overall mission by the engineering technician is 0.28.

4.9 Error-Cause Removal Program

This method was developed specifically to reduce human errors in production operations and its emphasis is on preventive measures rather than merely on remedial ones. Error-cause removal program (ECRP) may be described simply as the production worker participation program for reducing human errors. Workers such as machinists, assembly personnel, maintenance workers, and inspection personnel participate in the program [30]. All of these workers are grouped into a number of teams and each team has a maximum of 12 workers and a coordinator. The team meetings are held on a regular basis, during which the workers present error-likely and error reports.

The recommendations of the teams are presented to the management for appropriate preventive or remedial actions. Normally, management and teams are assisted by various types of specialists including human factors specialists.

Nonetheless, following are the basic elements of the ECRP [3,30]:

- All individuals involved with ECRP are educated in regard to its usefulness.
- Production workers report and determine errors and error occurrence-likely scenarios and put forward appropriate design solutions to eradicate error causes.
- All team coordinators and workers are trained in data collection and analysis methods.
- Human factors and other specialists evaluate proposed design-related solutions in regard to cost.
- Management recognizes the efforts of production workers and implements the most promising proposed design solutions.
- Human factors and other specialists, with the aid of the ECRP inputs, evaluate the effects of the changes made to the production process.

Additional information on the method is available in Refs. [3,30].

4.10 Markov Method

This is a widely used method to perform various types of reliability analysis of engineering systems. The method is named after a Russian mathematician, Andrei A. Markov (1856–1922). The method also can be used to perform various types of safety and human error analysis in engineering systems.

The following three assumptions are associated with the Markov method [1,31]:

- All occurrences are independent of each other.
- The transitional probability from one system state to another in the finite time interval Δt is given by $\mu \Delta t$, where μ is the transition rate (e.g., failure rate) from one system state to another.

- The probability of more than one transition occurrence in the finite time interval Δt from one system state to another is very small or negligible (e.g., $(\mu\Delta t)(\mu\Delta t)\to 0$).

The following example demonstrates the application of the method.

4.10.1 Markov Method Example

Example 4.3

Assume that an engineering system can either fail safely or unsafely due to a human error. The system safe failure rate is λ_1 and its unsafe failure rate due to a human error is λ_2. The system state space diagram is shown in Figure 4.6. The numerals in boxes denote system states. Develop expressions for the system state probabilities and mean time to failure ($MTTF_{es}$) by using the Markov method and assuming that the system failures occur independently and its failure rates are constant.

By using the Markov method, we write down the following equations for the system states 0, 1, and 2, respectively, shown in Figure 4.6.

$$P_0(t+\Delta t) = P_0(t)\,(1-\lambda_1\Delta t)\,(1-\lambda_2\Delta t) \tag{4.5}$$

$$P_1(t+\Delta t) = P_1(t)\,(1-0\Delta t) + P_0(t)\,\lambda_1\Delta t \tag{4.6}$$

$$P_2(t+\Delta t) = P_2(t)\,(1-0\Delta t) + P_0(t)\,\lambda_2\Delta t \tag{4.7}$$

where

t = time

$P_0(t+\Delta t)$ = the probability of the engineering system being in operating state 0 at time $(t+\Delta t)$

$P_1(t+\Delta t)$ = the probability of the engineering system being in safely failed state 1 at time $(t+\Delta t)$

$P_2(t+\Delta t)$ = the probability of the engineering system being in unsafely failed (due to a human error) state 2 at time $(t+\Delta t)$

$P_i(t)$ = the probability that the engineering system is in state i at time t, for $i = 0$ (operating normally), $i = 1$ (failed safely), and $i = 2$ (failed unsafely due to a human error)

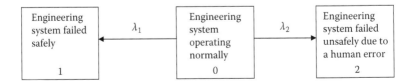

FIGURE 4.6
Engineering system state space diagram.

$\lambda_1 \Delta t$ = the probability of safe engineering system failure in finite time interval Δt

$(1 - \lambda_1 \Delta t)$ = the probability of no safe engineering system failure in finite time interval Δt

$\lambda_2 \Delta t$ = the probability of unsafe engineering system failure due to a human error in finite time interval Δt

From equation (4.5), we obtain:

$$P_0(t + \Delta t) = P_0(t) [1 - \lambda_1 \Delta t - \lambda_2 \Delta t + (\lambda_1 \Delta t)(\lambda_2 \Delta t)] \tag{4.8}$$

Because $(\lambda_1 \Delta t)(\lambda_2 \Delta t) \to 0$, equation (4.8) reduces to:

$$P_0(t + \Delta t) = P_0(t) [1 - \lambda_1 \Delta t - \lambda_2 \Delta t] \tag{4.9}$$

From equation (4.9), we write:

$$\lim_{\Delta t \to 0} it \frac{P_0(t + \Delta t) - P_0(t)}{\Delta t} = -P_0(t)\lambda_1 - P_0(t)\lambda_2 \tag{4.10}$$

Thus, from equation (4.10), we get:

$$\frac{dP_0(t)}{dt} + P_0(t)(\lambda_1 + \lambda_2) = 0 \tag{4.11}$$

Similarly, using equation (4.6) and equation (4.7), we obtain the following equation (4.12) and equation (4.13), respectively:

$$\frac{dP_1(t)}{dt} - P_0(t)\lambda_1 = 0 \tag{4.12}$$

$$\frac{dP_2(t)}{dt} - P_0(t)\lambda_2 = 0 \tag{4.13}$$

At time $t = 0$, $P_0(0) = 1$, $P_1(0) = 0$, and $P_2(0) = 0$.

By solving equation (4.11) to equation (4.13), we obtain:

$$P_0(t) = e^{-(\lambda_1 + \lambda_2)t} \tag{4.14}$$

$$P_1(t) = \frac{\lambda_1}{\lambda_1 + \lambda_2} \left[1 - e^{-(\lambda_1 + \lambda_2)t} \right] \tag{4.15}$$

$$P_2(t) = \frac{\lambda_2}{\lambda_1 + \lambda_2} \left[1 - e^{-(\lambda_1 + \lambda_2)t} \right] \tag{4.16}$$

By integrating equation (4.14) over the interval $[0, \infty]$, we obtain the following expression for the engineering system mean time to failure [1]:

$$MTTF_{es} = \int_0^\infty e^{-(\lambda_1+\lambda_2)t}dt$$

$$= \frac{1}{\lambda_1 + \lambda_2}$$

(4.17)

where
$MTTF_{es}$ = the engineering system mean time to failure

Example 4.4

Assume that the engineering system unsafe failure rate due to a human error is 0.0004 failures/hour and its safe failure rate is 0.0001 failures/hour. Calculate the engineering system safely failing probability during a 100-hour mission and its mean time to failure.

By substituting the given data values into equation (4.15), we get:

$$P_1(100) = \frac{0.0001}{0.0001 + 0.0004}\left[1 - e^{-(0.0001+0.0004)(100)}\right]$$

$$= 0.0097$$

Similarly, by substituting the given data values into equation (4.17), we obtain:

$$MTTF_{es} = \frac{1}{0.0001 + 0.0004}$$

$$= 2,000 \text{ hours}$$

Thus, the engineering system safely failing probability and mean time to failure are 0.0097 and 2,000 hours, respectively.

4.11 Fault Tree Analysis

This is a widely used method to evaluate engineering systems from the reliability aspect during their design and development, particularly in the area of nuclear power generation. The method was developed in the early 1960s at the Bell Telephone Laboratories to perform reliability analysis of the Minuteman Launch Control System [1,9]. A fault tree may simply be described as a logical representation of the relationship of basic or primary events that lead to the occurrence of a specified undesirable event known as the "top event" and is depicted using a tree structure with AND, OR, etc., logic gates.

Although there could be many purposes in performing fault tree analysis (FTA), some of the main ones are identifying critical areas and cost effective improvements, understanding the level of protection that the design concept provides against failures, understanding the functional relationship of system failures, and meeting jurisdictional requirements. FTA also can be used to perform various types of safety and human error analysis in engineering systems.

FTA starts by identifying an undesirable event, known as *top event*, associated with a system. Fault events that can cause the occurrence of the top event are connected and generated by logic operators, such as AND and OR. The AND gate provides a True output (i.e., fault) if all the inputs (i.e., faults) are true. Similarly, the OR gate provides a True output (i.e., fault) if one or more inputs (i.e., faults) are true.

A fault tree's construction proceeds by generating fault events in a successive manner until the fault events need not be developed any further. These fault events are called primary events. During the fault tree construction process, one successively asks the question: "How could this fault event occur?"

Four basic symbols used in constructing fault trees are shown in Figure 4.7 [1,9].

Each one of these four symbols is described below (it is to be noted that AND and OR gates are described again for the sake of clarity).

- **AND gate:** It denotes that an output event (fault) occurs only if all of the input events (faults) occur.

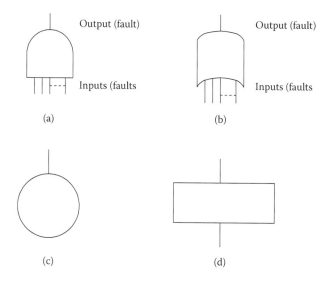

FIGURE 4.7
Fault tree symbols: (a) AND gate, (b) OR gate, (c) circle, (d) rectangle.

- **OR gate:** It denotes that an output event (fault) occurs if one or more of the input events (faults) occur.
- **Circle:** It represents a basic fault event (e.g., failure of an elementary component). The event's parameters are probability of occurrence, failure, and repair rates (the values of these parameters are usually obtained from empirical data).
- **Rectangle:** It represents a fault event that occurs from the logical combination of fault events through the input of a logic gate, such as OR and AND.

Information on other symbols used in performing FTA is available in Refs. [1,9].

Example 4.5

A windowless room has two light bulbs (i.e., X and Y) and one switch. The switch can fail to close. Develop a fault tree for the undesirable event (i.e., top fault event) "dark room" using Figure 4.7 symbols.

In this case, the room can only be dark if there is no incoming electricity, both the light bulbs burn out, or the switch fails to close. A fault tree for the example is shown in Figure 4.8.

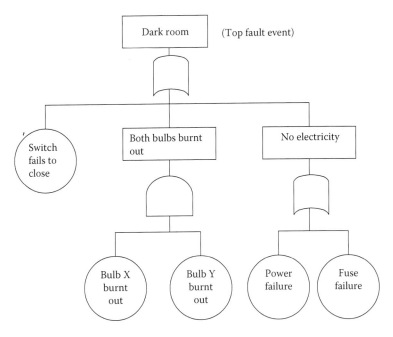

FIGURE 4.8
A fault tree for the top fault event: dark room.

4.11.1 Fault Tree Probability Evaluation

Under certain scenarios, it may be necessary to predict the occurrence probability of a certain event (e.g., unsafe failure of an engineering system due to human error). Before this could be achieved by using the FTA method, the determination of the occurrence probability of output fault events of logic gates is required.

Thus, the occurrence probability of the output fault event of an OR gate is expressed by [1,9]:

$$P_{og}(X) = 1 - \prod_{i=1}^{n} (1 - P(X_i)) \tag{4.18}$$

where

$P_{og}(X)$ = the occurrence probability of OR gate's output fault event X
n = the number of input fault events
$P(X_i)$ = the occurrence probability of input fault event X_i, for i = 1, 2, 3, ..., n.

Similarly, the occurrence probability of the output fault event of an AND gate is given by:

$$P_{ag}(X) = \prod_{i=1}^{n} P(X_i) \tag{4.19}$$

where

$P_{ag}(X)$ = the occurrence probability of AND gate's output fault event X

Example 4.6

Assume that the probabilities of occurrence of fault events—bulb X burnt out, bulb Y burnt out, switch fails to close, power failure, and fuse failure in Figure 4.8—are 0.09, 0.08, 0.07, 0.06, and 0.05, respectively. Using equation (4.18) and equation (4.19), calculate the probability of occurrence of the top fault event "dark room" and redraw the Figure 4.8 diagram with the calculated and given data values.

By substituting the given data values into equation (4.18), the probability of occurrence of the event "no electricity:"

$$P_{ne} = 0.06 + 0.05 - (0.06)(0.05) = 0.107$$

where

P_{ne} = the probability of occurrence of the event "no electricity"

Similarly, by inserting the specified data values into equation (4.19), the probability of occurrence of the event "both bulbs burnt out" is

$$P_{bb} = (0.09)(0.08)$$
$$= 0.0072$$

where

P_{bb} = the probability of occurrence of the event "both bulbs burnt out"

By substituting the above calculated values and the given data value into equation (4.18), the probability of occurrence of the top fault event "dark room" is

$$P_{dr} = (1 - (1 - 0.107)(1 - 0.0072)(1 - 0.07)$$
$$= 0.1755$$

Thus, the probability of occurrence of the top fault event "dark room" is 0.1755. Figure 4.8 diagram with the calculated and given data values is shown in Figure 4.9.

4.11.2 Fault Tree Analysis Benefits and Drawbacks

There are many benefits and drawbacks of the fault tree analysis. Some of the benefits include [1,9]:

- A graphic aid for management.
- Useful to provide insight into the system behavior and to identify failures deductively.

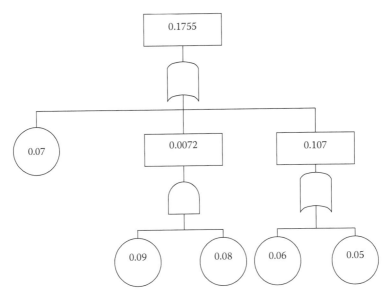

FIGURE 4.9
A fault tree with the calculated and given event occurrence probability values.

- Useful because it requires the analyst to understand thoroughly the system under consideration prior to starting the analysis.
- Useful to handle complex systems more easily.
- Useful because it allows concentration on one specific failure at a time.
- Useful to provide options for management and others to conduct either qualitative or quantitative analysis.

In contrast, some of the drawbacks are (1) a costly approach, (2) a time consuming method, (3) the end results are difficult to check, and (4) it considers parts in either working or failed state. More specifically, the partial failure states of the parts are difficult to handle.

Additional information on FTA is available in Refs. [1,9].

Problems

1. Describe interface safety analysis (ISA).
2. What are the advantages and disadvantages of root cause analysis?
3. Compare hazards and operability analysis with preliminary hazard analysis.
4. What is a technic of operations review?
5. Describe failure modes and effect analysis (FMEA).
6. What is the difference between FMEA and FMECA?
7. Assume that a safety professional performs four independent tasks: A, B, C, and D. Each of these four tasks can be performed either correctly or incorrectly and task A is performed before task B, task B before task C, and task C before task D. Develop a probability tree and obtain an expression for the probability of not successfully accomplishing the overall mission by the safety professional.
8. Prove that the sum of equation (4.14), equation (4.15), and equation (4.16) is equal to unity.
9. What are the advantages of FMEA?
10. Compare FMEA with FTA.

References

1. Dhillon, B. S., *Design Reliability: Fundamentals and Applications*, CRC Press, Boca Raton, FL, 1999.

2. Dhillon, B. S., *Engineering Safety: Fundamentals, Techniques, and Applications*, World Scientific Publishing, River Edge, NJ, 2003.
3. Dhillon, B. S., *Human Reliability: With Human Factors*, Pergamon Press, New York, 1986.
4. Dhillon, B. S., *Human Reliability, Error, and Human Factors in Engineering Maintenance: with Reference to Aviation and Power Generation*, CRC Press, Boca Raton, FL, 2009.
5. Coutinho, W. E., *Failure Effect Analysis*, Trans. New York Academy of Sciences, Series II, 1963–1964, pp. 564–584.
6. Palady, P., *Failure Modes and Effects Analysis*, PT Publications, West Palm Beach, FL, 1995.
7. Dhillon, B. S., *Patient Safety: An Engineering Approach*, CRC Press, Boca Raton, FL, 2012.
8. *Fault Tree Handbook*, Report No. NUREG-0492, United States Nuclear Regulatory Commission, Washington, D.C., 1981.
9. Dhillon, B. S., Singh, C., *Engineering Reliability: New Techniques and Applications*, John Wiley & Sons, New York, 1981.
10. Hammer, W., *Product Safety Management and Engineering*, Prentice Hall, Inc., Englewood Cliffs, NJ, 1980.
11. Goetsch, D. L., *Occupational Safety and Health*, Prentice Hall, Englewood Cliffs, NJ, 1996.
12. Hallock, R. G., Technique of Operations Review Analysis: Determine Cause of Accident/Incident, *Safety and Health*, Vol. 60, No. 8, 1991, pp. 38–39.
13. Busse, D. K., Wright, D. J., *Classification and Analysis of Incidents in Complex, Medical Environments*, Report, 2000. Available from the Intensive Care Unit, Western General Hospital, Edinburgh, U.K.
14. Latino, R. J., Automating Root Cause Analysis, in *Error Reduction in Health Care*, eds. P. L. Spath, John Wiley & Sons, New York, 2000, pp. 155–164.
15. Burke, A., *Root Cause Analysis*, Report, 2000. Available from the Wild Iris Medical Education, P.O. Box 257, Comptche, CA.
16. Dhillon, B. S., *Human Reliability and Error in Medical System*, World Scientific Publishing, River Edge, NJ, 2003.
17. Wald, H., Shojania, K. G., Root Cause Analysis, in *Making Health Care Safer: A Critical Analysis of Patient Safety Practices*, ed. A. J. Markowitz, Report No. 43, Agency for Health Care Research and Quality, U.S. Department of Health and Human Services, Rockville, MD, 2001, Chap. 5, pp. 1–7.
18. *Guidelines for Hazard Evaluation Procedures*, American Institute of Chemical Engineers, New York, 1985.
19. Dhillon, B. S., *Reliability, Quality and Safety for Engineers*, CRC Press, Boca Raton, FL, 2005.
20. Gloss, D. S., Wardle, M. G., *Introduction to Safety Engineering*, John Wiley & Sons, New York, 1984.
21. Roland, H. E., Moriarty, B., *System Safety Engineering and Management*, John Wiley & Sons, New York, 1983.
22. *Risk Analysis Requirements and Guidelines*, Report No. CAN/CSA-Q634-91, prepared by the Canadian Standards Association, 1991. Available from Canadian Standards Association, 178 Rexdale Blvd., Rexdale, Ontario, Canada.
23. Bahr, N. J., *System Safety Engineering and Risk Assessment: A Practical Approach*, CRC Press, Boca Raton, FL, 1997.

24. Omdahl, T. P., Ed., *Reliability, Availability, and Maintainability (RAM) Dictionary*, American Society for Quality Control (ASQC) Press, Milwaukee, WI, 1988.
25. MIL-F-18372 (Aer), *General Specification for Design, Installation, and Test of Aircraft Flight Control Systems*, Bureau of Naval Weapons, U.S. Department of the Navy, Washington, D.C.
26. MIL-STD-1629, *Procedures for Performing a Failure Mode, Effects, and Criticality Analysis*, U.S. Department of Defense, Washington, D.C., 1980.
27. Jordan, W. E., Failure Modes, Effects, and Criticality Analyses, paper presented at the Proceedings of the Annual Reliability and Maintainability Symposium, 1972, pp. 30–37.
28. McDermott, R. E., Mikulak, R. J., Beauregard, M. R., *The Basic of FMEA*, Quality Resources, New York, 1996.
29. Swain, A. D., *A Method for Performing a Human Factors Reliability Analysis*, Report No. SCR-685, Sandia Corporation, Albuquerque, NM, August 1963.
30. Swain, A. D., An Error-Cause Removal Program for Industry, *Human Factors*, Vol. 12, 1973, pp. 207–221.
31. Shooman, M. L., *Probabilistic Reliability: An Engineering Approach*, McGraw-Hill Book Company, New York, 1968.

5

Transportation Systems Safety

5.1 Introduction

Over the years, the safety of transportation systems has become an important issue. For example, in 1990, there were around 1 million traffic deaths and about 40 million traffic injuries worldwide and as per the projection of World Health Organization (WHO), the worldwide deaths from accidents will increase to around 2.3 million by 2020 [1,2].

For more than a century, railway safety has been an important issue in the United States. In 1893, the United States Congress passed the Federal Railway Safety Appliance Act. The Act instituted mandatory requirements for air brake systems and automatic couplers, and standardization of the specifications and locations for all appliances. Needless to say, actions such as this have helped to improve rail safety quite dramatically in the United States.

Each year, thousands of people die due to truck and bus safety-related problems. For example, in 2003, out of 42,643 traffic crash fatalities in the United States, 4986 involved large trucks [3,4]. The annual costs of large truck crashes and bus crashes for the period 1997 to 1999 in the United States were $19.6 billion and $0.7 billion, respectively [3].

Over the years, airline and ship safety has been a pressing issue, and various types of measures have been taken for its improvement. For example, in the area of civil aviation, the U.S. Congress passed the Air Commerce Act in 1926 [5,6]. The Act required the establishment of appropriate safety rules, the examination and licensing of aircraft and pilots, and proper investigation of accidents. Needless to say, although the safety in the area of aviation and sea transportation systems has improved quite dramatically over the years, it is still an important issue.

This chapter presents various important aspects of transportation systems safety.

5.2 Examples of Rail Accidents and Their Causes

Over the years, there have been many rail accidents around the world due to various causes. Some examples of these accidents include:

- In 1864, a passenger train in Ballinasloe, Ireland, derailed due to excess speed on a poor track and caused 2 fatalities and 34 injuries [7].
- In 1914, in Whangamarino, New Zealand, a Wellington to Auckland express train rear-ended a northbound freight train after it passed a faulty semaphore signal that wrongly displayed clear instead of danger and caused three fatalities and five serious injuries [8].
- In 1943, in Hyde, New Zealand, a Cromwell to Dunedin passenger train derailed on a curve due to excessive speed because of an intoxicated driver and caused 21 fatalities and 47 injuries [9].
- In 1949, in Donegal, Ireland, a passenger train left the station without train staff and collided head-on with a freight train and caused three fatalities and unknown number of injuries [7].
- In 1957, a Dundrum, Ireland, a passenger train, delayed by a cow on the line, was struck from behind by another passenger train mistakenly signaled into the station and caused one fatality and four injuries [10,11].
- In 1999, in Waipahi, New Zealand, a northbound Main South Line express freight train collided with a stationary southbound freight train due to misunderstanding of track warrant conditions by both train drivers and caused one fatality and one serious injury [9].
- In 2002, near Crescent City, Florida, USA, an Amtrak auto-train derailed due to malfunctioning brakes and poor track maintenance and caused 4 fatalities and 142 injuries [12].
- In 2004, in Macdona, Texas, USA, a Union Pacific Railway train failed to stop at a signal and collided with another train and caused 3 fatalities and 51 injuries [13].

5.3 Classifications of Rail Accidents by Causes and Effects

Rail accidents by causes can be grouped under many different classifications. The six common classifications are shown in Figure 5.1 [14–16]:

The "drivers' errors" classification includes items such as excessive speed, passing signals at danger, and engine mishandling. The "signalmen's errors" classification includes items such as allowing two trains into

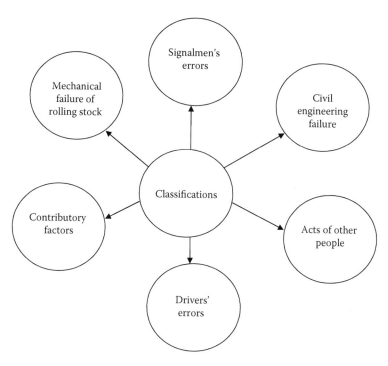

FIGURE 5.1
Classifications of rail accidents by causes.

the same occupied block section and wrong operation of signals, points, or token equipment. The "civil engineering failure" classification includes items such as tunnel and bridge collapses and track (permanent way) faults.

The "acts of other people" classification includes items such as the acts of other railway personnel (e.g., porters, shunters (worker who couples and uncouples cars)) and of nonrailway personnel (i.e., vandalism, terrorism, and accidental damage). The "contributory factors" classification includes items such as effectiveness of brakes, poor track or junction layout, rolling stock strength, and inadequate rules. Finally, the "mechanical failure of rolling stock" classification includes items such as poor design and maintenance.

Three commonly proposed classifications of the rail accidents by effects are collisions, derailments, and other [14–16]. The "collisions" classification includes items such as near collisions, collisions with buffer stops, head-on collisions, and obstructions on the line/track (i.e., landslides, avalanches, road vehicles, etc.). The "derailments" classification includes plain track, curves, and junctions. Finally, the "other" classification includes items such as fires and explosions, collisions with people on tracks, and falls from trains.

5.4 Railroad Tank Car Safety

Railroad tank cars are used to transport gases and liquids from one point to another, and their contents are corrosive, flammable, or pose other hazards if released accidently. During the period 1965 to 1980 in the United States, tank car accidents caused over 40 fatalities and accidental releases occur roughly once out of every 1000 shipments and result in about 1000 releases per year [17].

In order to ensure tank car safety, in the 1990 Hazardous Materials Transportation Uniform Safety Act, the United States Congress called for an examination of the tank car design process and an assessment of whether head shields should be made mandatory on all types of railroad tank cars that transport hazardous materials [17]. The Transportation Research Board (TRB), in order to address these two issues, formed a committee of experts in areas such as railroad operations and labor, tank car design, transportation and hazardous materials safety, chemical shipping, and transportation economics and regulations.

After examining railroad tank car incident-related data, the committee made the following three recommendations [17]:

1. Improve cooperation between the Department of Transportation and the industrial sector to highlight critical safety-related needs and take appropriate action to achieve them.
2. Improve the implementation of industry design approval and certification function and all federal oversight processes and procedures.
3. Improve the information and criteria used for assessing the safety-related performance of tank car design types and for assigning materials to tank cars.

5.5 Light-Rail Transit System Safety Issues

Currently, there are about 20 cities both in the United States and Canada that use light-rail transit systems. Over the years, there have been many accidents involving light-rail transit systems resulting in fatalities and injuries. For example, during the three year period following the opening of the Los Angeles' Metro Blue Line (MBL), there were a total of 158 train–vehicle and 24 train–pedestrian accidents/incidents causing 16 fatalities and many injuries [18].

Some of the safety issues involving light-rail transit system operations on city streets and on reserved rights of way with at-grade crossings include [18]:

- Traffic queues blocking designated crossing points.
- Crossing equipment malfunction.

- Pedestrians conflict at station areas and designated crossing points.
- Light-rail vehicles blocking road/street and pedestrians' cross walk areas at designated crossing points.
- Confusion of motorists over light-rail transit signals, traffic signals, and signage at intersection points.
- Vehicles turning from roads/streets that run exactly parallel to the rail tracks.
- Motorists' disobedience in regard to traffic laws.
- Motor vehicles making left or U-turns in front of oncoming rail vehicles or stopping on rail tracks.

5.6 Truck Safety-Related Facts and Figures

Some of the truck safety-related facts and figures include:

- In 1980, 1986, 1989, 1992, 1995, 1997, and 2000 there were around 5400, 5100, 5000, 4000, 4500, 4900, and 5000 truck-related fatal crashes in the United States, respectively [3].
- In 1993, approximately 80% of the truck accidents in the United States occurred with no adverse weather conditions [19].
- In 2003, in the United States, out of 4986 fatalities that occurred from crashes involving large trucks, 78% were occupants of another vehicle, 14% were occupants of large trucks, and 8% were nonoccupants [3,20].
- As per Ref. [19], approximately 65% of large truck crash deaths in the United States occur on major roads.
- During the period from 1976 to 1987, truck occupants' fatalities decreased from 1130 in 1976 to 852 in 1987 in the United States [21].
- In 1993, in the United States, about 4500 trucks were involved in accidents in which at least one death occurred [19].
- During the period from 1976–1987, in the United States, the large trucks' fatal crash rate declined by 20% [3].

5.7 Truck and Bus Safety-Related Issues

Over the years, many studies have been conducted to identify truck and bus safety-related issues. Some of the important ones are shown in Figure 5.2 [22].

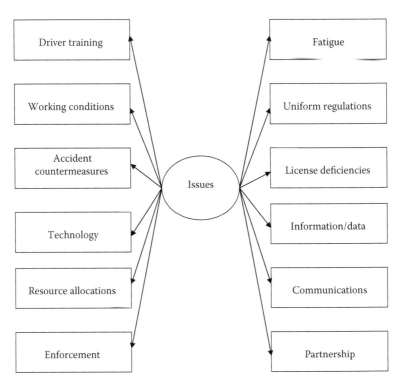

FIGURE 5.2
Truck and bus safety-related issues.

The issue "driver training" is concerned with the need for continuing and better education for all drivers (i.e., private and commercial). The issue "fatigue" is concerned with driving, scheduling, unloading, and road conditions that induce fatigue, in addition to a lack of appropriate places for rest and hours-of-service violations. The issue "license deficiencies" is concerned with the review of testing procedures followed for licenses of commercial drivers.

The issue "communications" is concerned with the development of an effective national motor-carrier safety marketing campaign as well as the expansion of education-related efforts to the public-at-large for sharing roads with large vehicles. The issue "information/data" is concerned with the shortage of information on heavy vehicle crashes and their causes. The issue "uniform regulations" is concerned with the lack of uniformity in safety-related procedures and regulations among the states and between Mexico and Canada, clearly indicating that safety issues do not receive the same level of priority in all jurisdictions. The issue "partnership" is concerned with better coordination and communication among highway users.

The issue "enforcement" is concerned with the need for better traffic enforcement, licensing and testing, and adjudication of highway user violations. The issue "technology" is concerned with the development and deployment of appropriate emerging and practically inclined technologies to improve safety. The issue "accident countermeasures" is concerned with the appropriate research efforts targeted for seeking and defining proactive and nonpunitive countermeasures to prevent accidents. The issue "working conditions" is concerned with the review of on-going industry practices and standards as they affect workload of drivers. Finally, the issue "resource allocations" is concerned with the priorities and allocation of all scarce resources through a better safety-management system that clearly gives safety top priority.

5.8 Commonly Cited Truck Safety-Related Problems and Recommendations for Improving Truck Safety

Although there are many truck safety-related problems, the most commonly cited ones include [19]:

- Head-on collisions
- Rollovers
- Angle impact
- Rear-end collisions
- Jackknifes
- Sideswipes

The problem "head-on collisions" accounts for approximately 24% of fatal truck involvements, 1.9% of tow-away involvements, 1.6% of injury involvements. The problem "rollovers" accounts for about 13.3% of fatal truck involvements, 10.8% of injury involvements, and 8.6% of tow-away involvements. The problem "angle impact" accounts for around 37.7% of tow-away involvements, 37% of injury involvements, and 32.5% of fatal truck involvements.

The problem "rear-end collisions" accounts for 30.6% of injury involvements, 26.9% of tow-away involvements, and 18.4% of fatal truck involvements. The problem "jackknifes" occurs when a multiunit vehicle (e.g., tractor-trailer) folds up like a pocket knife and accounts for approximately 8.4% of tow-away involvements, 8.3% of fatal truck involvements, and 5.5% of injury involvements. Finally, the problem "sideswipes" accounts for roughly 9.7% of tow-away involvements and 4.1% of fatal truck involvements.

In 1995, the attendees of a conference on "Truck Safety: Perceptions and Reality" made many recommendations on the issues shown in Figure 5.3 to improve truck safety [23,24].

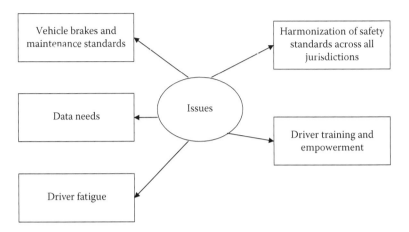

FIGURE 5.3
Truck safety improvement recommendations' issues.

The following four recommendations were on the issue of vehicle brakes and maintenance standards [24]:

1. Invoke appropriate penalties for those trucking organizations or companies that regularly fail to meet required inspection standards.
2. Train and certify truck drivers to adjust vehicle brakes appropriately as part of their licensing requirements and training program.
3. Implement an effective safety rating system.
4. Equip trucks with essential onboard devices/signals for indicating when brakes need servicing and adjustment.

The following four recommendations were on the issue of driver training and empowerment [24]:

1. Devise and enforce appropriate regulations to ensure that truck drivers are not unfairly dismissed when they refuse to drive in unsafe conditions.
2. Establish appropriate driver training and retraining programs that focus on safety (e.g., teaching drivers how to inspect the vehicle by using the latest technology) and to take necessary measures to reduce accident risk.
3. Enact an accreditation of all driver training schools for ensuring that they uniformly meet required standards in all jurisdictions.
4. Implement a graduated licensing scheme that reflects the need of a variety of trucking vehicles.

The following four recommendations were on the issue of data needs [24]:

1. Establish a North American Truck Safety data center.
2. Identify and share currently available truck accident and exposure-related data.
3. Improve reliability of police accident reports through better police training for reporting and collecting reliable data on accident consequences and causes.
4. Standardize accident reporting forms used by police in all jurisdictions.

The following three recommendations were on the issue of driver fatigue [24]:

1. Harmonize applicable standards across different jurisdictions.
2. Set tolerance levels for fatigue and accident risk and devise new standards that incorporate these levels.
3. Develop a comprehensive approach for identifying the incidence of fatigue of truck drivers that clearly takes in consideration different types of fatigue and driving-related needs.

Finally, the following two recommendations were on the issue of harmonization of safety standards across all jurisdictions [24]:

1. Form a committee of industry and government representatives for exploring avenues for cooperative efforts to develop uniform truck safety standards.
2. Establish an appropriate agency to collect and disseminate safety information to concerned parties.

5.9 Transit Bus Safety and Important Design-Related Safety Feature Areas

Although buses are considered as one of the safest modes of transportation, during the period from 1999 to 2003 in the United States, there was an annual average of 40 bus occupant fatalities and 18,430 injuries [25]. Furthermore, with respect to two-vehicle crashes, there was an annual average of 11 bus occupant fatalities, while there was an annual average of 162 fatalities for occupants of other vehicles (i.e., 102 occupants in passenger cars, 49 in light trucks, 9 on motorcycles, and 2 in large trucks) [25].

During the period from 1999 to 2001, in the United States, there was an annual average of 111 transit buses involved in fatal accidents [25,26]. Additional transit bus safety-related information is available in Refs. [25,27].

Over the years, many design-related safety feature areas for improving transit bus safety have been identified by the professionals working in the area. The important ones include [28]:

- Better external designs that remove potentially dangerous items such as handholds, protrusions, and footholds
- Wide doors, low floors, and energy-absorbing sidewalls and bumpers
- Better lighting and visibility for both drivers and passengers
- Interior designs of the transit bus based on considerations that feature selective padding and removal of protrusions considered dangerous

5.10 World Airline Accident Analysis and United States Airline-Related Fatalities

Airlines have become a widely used mode of transportation around the globe. Currently, over 16,000 jet aircraft are being used throughout the world, with over 17 million departures [5]. A study of worldwide scheduled commercial jet operations during the period from 1959 to 2001 reported that there were 1307 accidents, causing 24,700 onboard deaths [5,29]. By the type of operation, three classifications of these 1307 accidents are shown in Figure 5.4 [5].

It is to be noted that the collective Canadian and United States element of these 1307 accidents was about 34% (i.e., 445 accidents), causing around 25% (i.e., 6077) of the 24,700 onboard deaths [5].

A study of the accident data for the period 1959 to 2001 revealed that the world commercial jet fleet accident rate (i.e., accidents per million departures) was fairly stable for the period from 1974 to 2001 [29]. Additional information on the topic is available in Refs. [5,29].

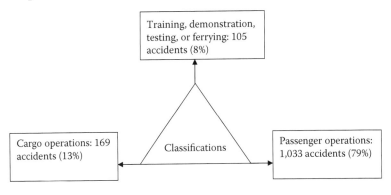

FIGURE 5.4
Classifications of 1307 commercial jet operation-related accidents.

TABLE 5.1

Accident Rates per Million Flight Departures
in the United States, 1989–1995

Year	Accident Rate (Per Million Flight Departures)
1989	0.37
1990	0.29
1991	0.33
1992	0.22
1993	0.28
1994	0.27
1995	0.40

The history of fatal airline accidents in the United States goes back to 1926 and 1927, when there were 24 fatal commercial airline accidents. In 1929, 61 people were killed in 51 airline accidents. This year still remains the worst year on record, with an accident rate of roughly 1 per million miles flown [6,30]. For the years 1983 through 1995, the number of fatalities due to commercial airline accidents in the United States were 8, 0, 486, 0, 212, 255, 259, 8, 40, 35, 0, 228, and 152, respectively [6]. The accident rates per million departures for the period 1989 to 1995 are presented in Table 5.1 [6].

In comparison to fatalities in other areas, the airline-related fatalities in the United States are extremely low. For example, in 1995 people were roughly 30 times more likely to drown and roughly 300 times more likely to die in a motor vehicle accident than to get killed in an airplane-related accident [6].

5.11 Causes of Airplane Crashes

There are many causes for the occurrence of airplane crashes. For example, a study of 19 major crashes (defined as one in which at least 10% of the airplane passengers die) of United States domestic jets occurring during the period from 1975 to 1994, highlighted eight main causes of these crashes [6,31]. These eight causes along with their corresponding number of crashes (in parentheses) are shown in Figure 5.5 [6,31].

5.12 Air Safety-Related Regulatory Bodies and Their Responsibilities

In the United States, there are two air safety-related regulatory bodies that serve as the public's watchdog for safety in the aviation industrial

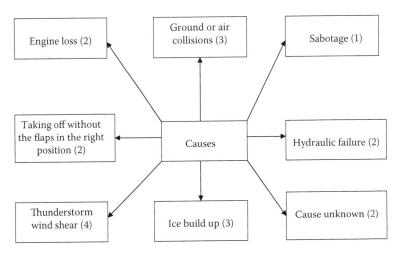

FIGURE 5.5
Main causes of major airplane crashes along with their corresponding number of crashes in parentheses.

sector. These bodies are the Federal Aviation Administration (FAA) and the National Transportation Safety Board (NTSB). The history of both the FAA and the NTSB goes back to 1940 when the Civil Aeronautics Authority was split into two organizations: the Civil Aeronautics Board (CAB) and the Civil Aeronautics Administration (CAA).

Since then, the CAB has evolved into the National Transportation Safety Board (NTSB) and the CAA into the Federal Aviation Administration (FAA) [6]. Nowadays, both the NTSB and the FAA are within the framework of the U.S. Department of Transportation [6].

The responsibilities of the NTSB are within and outside the aviation industry. More specifically, the NTSB also is responsible for investigating significant accidents occurring in other modes of transportation, such as railroad and marine, in addition to investigating accidents in the aviation industry. The main responsibilities of the NTSB include [6]:

- Conduct special studies on issues concerning transportation safety
- Serve as the "court of appeals" for FAA-related matters
- Maintain the government database on accidents occurring in the aviation industry
- Issue safety-related recommendations to prevent potential accidents

Similarly the main responsibilities of the FAA include [6,32]:

- Develop airline safety regulations
- Develop, operate, and maintain the nation's air control system

- Establish minimum standards for crew training
- Develop appropriate operational requirements for airlines
- Review the design, manufacture, and maintenance of aircraft-related equipment
- Conduct safety research and development

5.13 Marine Accidents

Over the years, many marine accidents have occurred around the world. Three of the more noteworthy of these accidents are described below.

5.13.1 The Derbyshire Accident

This accident is concerned with a large bulk carrier ship named *Derbyshire*. The ship disappeared during a typhoon in the Pacific Ocean on September 9, 1980, en route to Kawasaki, Japan, carrying a cargo of iron ore concentrates. The accident resulted in 44 fatalities (i.e., 42 crew members and 2 wives) [32,33].

The ship was designed in compliance with freeboard and hatch cover strengths as stated in the 1968 regulations of the UK Government [34]. All in all, the minimum requirements of hatch cover strength for forward hatch covers for bulk carrier ships of about same size to the *Derbyshire* are considered seriously deficient in regard to the current acceptable safety levels [32,33].

5.13.2 The Herald of Free Enterprise Accident

This accident concerned a passenger ship named *Herald of Free Enterprise*. On March 6, 1987, the ship left Zeebrugge Harbor, Belgium, and about five minutes after its departure, it capsized and caused over 180 fatalities [33,35]. The ship's capsizing was due to a combination of adverse factors, including the vessel speed, the bow door being left open, and trim by the bow.

The subsequent public inquiry into the *Herald of Free Enterprise* disaster is considered as an important milestone in the history of ship safety in the United Kingdom. It resulted in a number of actions including the introduction of the International Safety Management (ISM) code for the safe operation of ships and for pollution prevention, the development of a formal safety assessment process in the shipping industry, and changes to marine safety rules and regulations [33].

5.13.3 The Estonia Accident

This accident is concerned with an Estonian-flagged roll-on-roll (RO-RO) passenger ferry named *Estonia*. On September 27, 1994, the ferry left Tallinn, the capital city of Estonia, for Stockholm, Sweden, carrying 989 people on board, and in the early hours of September 28, it sank in the Baltic Sea [32,33]. The accident caused 852 fatalities.

A subsequent investigation into the disaster reported that the bow visor locks of the ferry were too weak due to their poor design and manufacture. All in all, these locks broke during bad weather and the visor fell off by pulling open the inner bow ramp causing the disaster [33,34].

5.14 Ship Port-Related Hazards

There are many ship port-related hazards and they may be grouped under eight classifications, as shown in Figure 5.6 [36].

The classification "loss of containment" contains those hazards that are concerned with the release and dispersion of dangerous substances. Two typical examples of these hazards are release of toxic material and release of flammables.

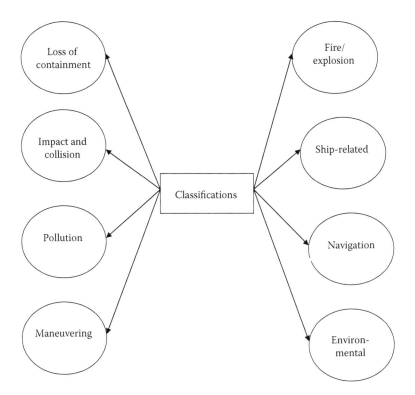

FIGURE 5.6
Classifications of ship port-related hazards.

The classification "pollution" contains those hazards that are concerned with the release of material that can cause damage to the environment. An example of these hazards is crude oil spills. The classification "navigation" contains those hazards that have potential for a deviation of the ship from its intended designated channel or route. Three examples of these hazards are pilot error, vessel not under command, and navigation error.

The classification "environmental" contains those hazards that occur when weather exceeds vessel design criteria or harbor operations criteria. Three examples of these hazards are strong currents, extreme weather, and winds exceeding port criteria. The classification "impact and collision" contains those hazards that are concerned with an interaction with a stationary or a moving object, or a collision with a vessel. Three examples of these hazards are berthing impacts, striking while at berth, and vessel collision. The classification "fire/explosion" contains those hazards that are concerned with explosion or fire in the cargo bay or on the vessel. Two examples of these hazards are fire in engine room and cargo tank fire/explosion.

The classification "maneuvering" contains those hazards that are concerned with failure to keep the vessel on the right track or to position the vessel as intended. Two examples of these hazards are berthing/unberthing error and fine-maneuvering error. Finally, the classification "ship-related" contains those hazards that are concerned with ship-specific equipment or operations. Four examples of these hazards are anchoring failure, flooding, loading/overloading, and mooring failure.

5.15 Ship Safety Assessment and Global Maritime Distress Safety System

In recent years, the assessment of ship safety has become an important issue. Risk and cost-benefit assessment methods are being used during the ship-safety decision-making process. The approach or procedure is called a formal safety assessment (FSA) and is made up of five steps, as shown in Figure 5.7 [32,33].

Step 1 is concerned with identifying hazards specific to a ship safety problem under review. A hazard is defined as a physical condition with a potential for damage to property, damage to the surrounding environment, human injury, or some combination of these. Step 2 is concerned with estimating risks and factors influencing the level of ship safety. Step 3 is concerned with proposing practical and effective risk-control options. The results of Step 2 are used to identify high-risk areas to propose necessary risk-control actions.

Step 4 is concerned with the identification of benefits from reduced risks and costs associated with the implementation of each identified risk-control

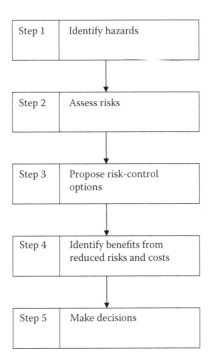

FIGURE 5.7
Formal safety assessment approach steps.

option. Finally, Step 5 is concerned with making appropriate decisions and providing necessary recommendations for ship safety-associated improvements.

The Global Maritime Distress Safety System (GMDSS) is based on a combination of satellite and terrestrial radio services and provides for fully automatic alerting and locating, thus completely eliminating the need for a radio operator to send an SOS/a Morse code distress signal (i.e., Mayday call). It may simply be characterized as an internationally agreed-upon set of safety-related procedures, communication protocols, and equipment types employed to increase safety and to make it easier for rescuing distressed boats, ships, and aircraft. Nonetheless, the main functions of the GMDSS include [32,37]:

- Maritime safety-related information broadcasts
- Search and rescue coordination
- General communications
- Bridge-to-bridge communications
- Locating (homing)
- Alerting (including position determination of the unit in distress)

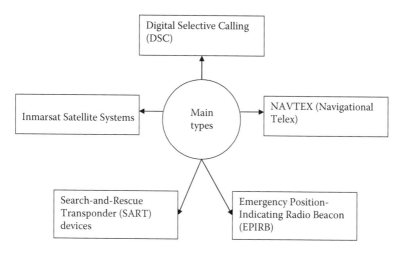

FIGURE 5.8
Main types of equipment used in the Global Maritime Distress Safety System (GMDSS).

Here, it is to be noted that GMDSS also provides redundant means of distress alerting and emergency power sources. Some of the main types of equipment/systems used in GMDSS are shown in Figure 5.8 [32,27].

Additional information on GMDSS is available in Refs. [32,37].

Problems

1. Write an essay on transportation systems safety.
2. Discuss at least five rail accidents and their causes.
3. Discuss railroad tank car safety.
4. Discuss important light-rail transit system safety issues.
5. List at least five facts and figures concerned with truck safety.
6. Briefly discuss at least 10 important truck and bus safety-related issues.
7. What are the main causes of major airplane crashes?
8. What are the main responsibilities of the Federal Aviation Administration (FAA) and the National Transportation Safety Board (NTSB)?
9. Discuss the following two marine accidents:
   a. *The Herald of Free Enterprise* accident
   b. The *Derbyshire* accident
10. Discuss ship port-related hazards.

References

1. Murray, C. J. L., Lopez, A. D., The Global Burden of Disease in 1990: Final Results and Their Sensitivity to Alternative Epidemiological Perspectives, Discount Rates, Age-Weights, Disability Weights, In *The Global Burden of Disease*, eds. C. J. L. Murray and A. D. Lopez, Harvard University Press, Cambridge, MA, 1996, pp. 15–24.
2. Freund, P. E. S., Martin, G. T., Speaking about Accidents: The Ideology of Auto Safety, *Health*, Vol. 1, No. 2, 1997, pp. 167–182.
3. *The Domain of Truck and Bus Safety Research*, Transportation Research Circular No. E-C117, Transportation Research Board, Washington, D.C., 2007.
4. Zaloshnja, E., Miller, T., *Revised Costs of Large Truck-and Bus-Involved Crashes*, Final Report, Contract No. DTMC75-01-P-00046, Federal Motor Carrier Safety Administration (FMCSA), Washington, D.C., 2002.
5. Wells, A. T., Rodrigues, C. C., *Commercial Aviation Safety*, McGraw Hill, New York, 2004.
6. Benowski, K., Safety in the Skies, *Quality Progress*, January 1997, pp. 25–35.
7. MacAongusa, B., *Broken Rails*, Currach Press, Dublin, Ireland, 2005.
8. Conly, G., Stewart, G., *Tragedy on the Track: Tangiwai and other New Zealand Railway Accidents*, Grantham House Publishing, Wellington, New Zealand, 1986.
9. Transportation Accident Investigation Commission (TAIC), Wellington, New Zealand. Available online at http://www.taic.org.nz/ (retrieved on July 26, 2011).
10. MacAongusa, B., *The Harcourt Street Line: Back on Track*, Currach Press, Dublin, Ireland, 2003.
11. Murray, D., Collision at Dundrum, *Journal of the Irish Railway Record Society*, Vol. 17, No. 116, 1991, pp. 434–441.
12. *Derailment of Amtrak Auto Train Po52-18 on the CSXT Railroad Near Crescent City, FL, April 18, 2002*, Report No. RAR-03/02, National Transportation Safety Board (NTSB), Washington, D.C., 2003.
13. Report No. 04-03, Macdona Accident, National Transportation Safety Board (NTSB), Washington, D.C., 2004.
14. Schneider, W., Mase, A., *Railway Accidents of Great Britain and Europe: Their Causes and Consequences*, David and Charles, Newton Abbot, U.K., 1970.
15. Rolt, L. T. C., *Red for Danger*, David and Charles, Newton Abbot, U.K., 1966.
16. Classification of Railway Accidents, Wikipedia, 2010. Available online at http://en.wikipedia.org/wiki/Classification_of_railway_accidents.
17. Ensuring Railroad Tank Car Safety, *TR News*, No. 176, January-February 1995, pp. 30–31.
18. Meadow, L., Los Angeles Metro Blue Line Light Rail Safety Issues, *Transportation Research Record*, Vol. 1433, 1994, pp. 123–133.
19. Cox, J., Tips on Truck Transportation, *American Papermaker*, Vol. 59, No. 3, 1996, pp. 50–53.
20. Report No. FMCSA-RI-04-033, *Large Truck Crash Facts 2003*, Federal Motor Carrier Safety Administration (FMCSA), Washington, D.C., 2005.
21. Sheiff, H. E., Status Report on Large-Truck Safety, *Transportation Quarterly*, Vol. 44, No. 1, 1990, pp. 37–50.

22. Hamilton, S., The Top Truck and Bus Safety Issues, *Public Roads*, Vol. 59, No. 1, 1995, p. 20.
23. Gillen, M., Baltz, D., Gassel, M., Kirsch, L., Vaccaro, D., Perceived Safety Climate, Job Demands, and Coworker Support among Union and Non-union Injured Construction Workers, *Journal of Safety Research*, Vol. 33, 2002, pp. 33–51.
24. Saccomanno, F. F., Craig, L., Shortreed, J. H., Truck Safety Issues and Recommendations: Results of the Conference on Truck Safety: Perceptions and Reality, *Canadian Journal of Civil Engineers*, Vol. 24, 1997, pp. 326–332.
25. Olivares, G., Yadav, V., *Mass Transit Bus-Vehicle Compatibility Evaluations during Frontal and Rear Collisions*, Proceedings of the 20th International Conference on the Enhanced Safety of Vehicles, 2007, pp. 1–13.
26. Matteson, A., Blower, D., Hershberger, D., Woodrooffe, J., *Buses Involved in Fatal Accidents: Fact Book 2001, Center for National Truck and Bus Statistics*, Transportation Research Institute, University of Michigan, Ann Arbor, MI, 2005.
27. Yang, C. Y. D., Trends in Transit Bus Accidents and Promising Collision Countermeasures, *Journal of Public Transportation*, Vol. 10, No. 3, 2007, pp. 119–136.
28. Mateyka, J. A., *Maintainability and Safety of Transit Buses*, Proceedings of the Annual Reliability and Maintainability Symposium, 1974, pp. 217–225.
29. *Statistical Summary of Commercial Jet Aircraft Accidents: Worldwide Operations 1959–2001*, Boeing Commercial Airplane Company, Seattle, WA, 2001.
30. *Aviation Accident Synopses and Statistics*, National Transportation Safety Board (NTSB), Washington, D.C., 2007. Available online at http://www.ntsb.gov/Aviation/Aviation.htm.
31. Beck, M., Hosenball, M., Hager, M., Springen, K., Rogers, P., Underwood, A., Glick, D., Stanger, T., How Safe Is This Flight? *Newsweek*, April 24, 1995, pp. 18–29.
32. Dhillon, B. S., *Transportation Systems Reliability and Safety*, CRS Press, Boca Raton, FL, 2011.
33. Wang, J., Maritime Risk Assessment and Its Current Status, *Quality and Reliability Engineering International*, Vol. 22, 2006, pp. 3–19.
34. Wang, J., A Brief Review of Marine and Offshore Safety Assessment, *Marine Technology*, Vol. 39, No. 2, 2002, pp. 77–85.
35. *Herald of Free Enterprise: Fatal Accident Investigation*, Report No. 8074, United Kingdom Department for Transport, Her Majesty's Stationery Office (HMSO), London, 1987.
36. Trbojevic, V. M., Carr, B. J., Risk Based Methodology for Safety Improvement in Ports, *Journal of Hazardous Materials*, Vol. 71, 2000, pp. 467–480.
37. *Global Maritimes Distress Safety System (GMDSS)*, Wikipedia, 2010. Available online at http://en.wikipedia.org/wiki/Global_Maritime_Distress_Safety-System.

6

Medical Systems Safety

6.1 Introduction

Each year billions of dollars are spent to produce various types of medical systems/devices for use in the area of healthcare around the globe. A medical system or device must not only be reliable, but also safe for users and patients. The problem of safety concerning humans is not new; it can be traced back to the ancient Babylonian ruler Hammurabi. In 2000 BCE, Hammurabi developed a code known as the "Code of Hammurabi" in regard to health and safety [1–3]. The Code contained clauses with respect to injuries and financial damages against those causing injury to others.

In modern times, the passage of the Occupational Safety and Health Act (OSHA) by the U.S. Congress in 1970 is considered to be an important milestone in regard to health and safety in the United States. Two other important milestones that are specifically concerned with medical devices in the United States are the Safe Medical Device Act (SMDA) in 1990 and the Medical Device Amendments of 1976.

This chapter presents various important aspects of medical systems safety.

6.2 Facts and Figures

Some of the facts and figures, directly or indirectly, concerned with medical systems safety include:

- In 1969, the special committee of the U.S. Department of Health, Education, and Welfare reported that over a period of 10 years, there were about 10,000 medical device-associated injuries and 731 caused fatalities [4,5].
- A drug overdose occurred due to wrong advice given by an artificial intelligence medical system [6].

- After examining a sample of 15,000 hospital products, Emergency Care Research Institute (ECRI) concluded that 4 to 6% of these products were dangerous enough to warrant immediate corrective measure [7].
- A patient fatality occurred due to radiation overdose involving a Therac radiation therapy device [8].
- As per Ref. [6], faulty software programs in heart pacemakers caused two fatalities.
- As per Ref. [9], over time, ECRI has received many reports concerning radiologic equipment failures that either caused or had the potential to result in serious patient injury or death.
- A five-year-old patient was crushed to death beneath the pedestal-style electric bed in which the child was placed after hospital admission [10].

6.3 Medical Device Safety versus Reliability and Medical Device Hardware and Software Safety

Although both safety and reliability are good things to which medical devices should aspire, time to time there is some confusion, particularly in industry, in regard to the difference between medical device reliability and safety. Nonetheless, it is to be noted that reliability and safety are quite distinct concepts and at times they can have rather conflicting concerns [11].

A safe medical device/system may simply be described as a device/system that does not cause too much risk to property, humans, or equipment [3]. In turn, risk is an undesirable event that can occur and is measured in regard to probability and severity. More simply, device/system safety is a concern with malfunctions or failures that introduce hazards and is expressed in terms of the level of risk, not in terms of satisfying stated requirements. On the other hand, a medical device/system reliability is the probability of success to satisfy its stated requirements.

Finally, it is to be noted that a medical device/system is still considered safe even if it frequently fails without causing any mishap. In contrast, if a device/system functions normally at all times, but regularly puts humans at risk, under this scenario the device/system is considered reliable but unsafe. Some examples of both these scenarios are available in Refs. [3,11].

Medical device/system hardware safety is important because parts, such as electronic parts, are quite vulnerable to factors such as electrical interferences and environment-related stresses. It simply means that each and every part in a medical device must be analyzed with care in regard to potential safety concerns and failures. For this very purpose, there are various approaches available to the involved analyst including failure modes and effect analysis (FMEA) and fault tree analysis (FTA). Subsequent to part analysis, methods, such as part derating, safety margin, and load protection, can be utilized to reduce the potential for the occurrence of failure of parts pinpointed as critical [3,11].

Safety of device software is equal in importance to the safety of the device hardware parts. However, it is added that software in and of itself is not really unsafe, but the physical systems/devices it may control can cause damage of varying degree. For example, an out of control software program can drive the gantry of a radiation therapy machine into a patient, or a hung software program may not only malfunction to stop a radiation exposure, but also deliver an overdose to a certain degree [11]. All in all, it is to be noted that the software safety problem in medical devices is quite serious, as pointed out in a U.S. Food and Drug Administration (FDA) "device recalls" study [3,11]. More specifically, the study performed over a five-year period (i.e., from 1983–1989) reported that 116 problems pertaining to software quality led to the recall of medical devices in the United States.

6.4 Types of Medical Device Safety and Essential Safety-Related Requirements for Medical Devices

Medical device safety may be categorized under the following three classifications or types [12]:

Classification I: Unconditional Safety. This type of safety is preferred over all other possibilities or types because it is most effective. However, it calls for eradication of all device-related risks through design. Furthermore, it is to be noted with care that the use of warnings complements satisfactory device design, but does not replace it.

Classification II: Conditional Safety. This type of safety is used in situations when unconditional safety cannot be realized. For example, in the case of an x-ray/laser surgical device, it is impossible to avoid dangerous radiation emissions. However, it is well within means to minimize risk with actions, such as incorporating a locking mechanism that allows device activation by authorized personnel only or limiting access to therapy rooms. Two examples of the indirect safety means are protective laser glasses and x-ray folding screens.

Classification III: Descriptive Safety. This type of safety is used in situations when it is impossible or inappropriate to provide safety through the above two means (i.e., conditional or unconditional). Nonetheless, descriptive safety in regard to maintenance, operation, mounting, connection, transport, and replacement may simply be statements, such as Handle with Care, Not for Explosive Zones, This Side Up, etc.

There are various types of, directly or indirectly, safety-related requirements placed by the government and other agencies on medical devices. These requirements may be grouped under the following three areas [12]:

- Safe design
- Safe function
- Sufficient information

The requirements belonging to the safe design area are mechanical hazard prevention, excessive heating prevention, care for environmental conditions, protection against radiation hazards, care for hygienic factors, protection against electrical shock, and proper material choice with respect to mechanical, chemical, and biological factors. Mechanical hazard prevention includes factors such as safe distances, stability of the device, and breaking strength. The excessive heating prevention-related mechanisms are temperature control, cooling, and effective design. The care for environmental conditions includes factors such as temperature, electromagnetic interactions, and humidity. The remaining requirements are considered self-explanatory, but the additional information on them is available in Ref. [12].

The components of the safe function group are warning for or prevention of hazardous outputs, reliability, and accuracy of measurements. Finally, the sufficient information group includes items such as effective labeling, instructions for use, packaging, and accompanying documentation.

6.5 Safety in Medical Device Life Cycle

Past experience indicates that, in order to have safe medical devices, safety has to be considered throughout their life cycle. Thus, the life cycle of a medical device may be divided into five phases, as shown in Figure 6.1 [3,13].

In the concept phase, past data and future technical-related projections become the basis for the device under consideration and safety-associated problems are identified and evaluated. The preliminary hazards analysis (PHA) method is a quite effective tool to identify hazards during this phase. At the end of this phase, some of the typical questions to ask in regard to device safety include [3,13]:

- Are all the basic safety design-related requirements for the phase in place so that the definition phase can be started?
- Are all the hazards identified and properly evaluated to develop hazard controls?
- Is the risk analysis initiated to develop mechanisms for hazard control?

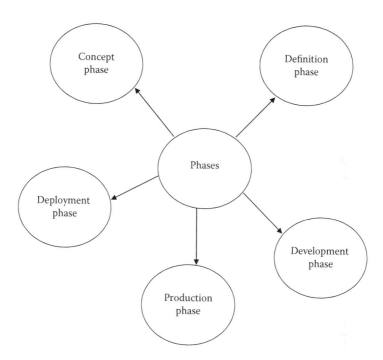

FIGURE 6.1
Medical device life cycle phases.

The main objective of the definition phase is to provide proper verification of the initial design and engineering concerned with the medical device under consideration. The results of the PHA are updated along with the initiation of subsystem hazard analysis and their ultimate integration into the overall device hazard analysis. Methods, such as fault hazard analysis and fault tree analysis (FTA), may be utilized for examining certain known hazards and their effects. All in all, the system definition will initially lead to the acceptability of a desirable general device design even though, because of the incompleteness of the design, not all associated hazards will be completely known.

During the development phase of the device, the efforts are directed on areas such as environmental impact, operational use, integrated logistics support, and producibility engineering. With the aid of prototype analysis and testing results, the comprehensive PHA is conducted to examine man–machine associated hazards, in addition to developing PHA further because of more completeness of the design of the device under consideration.

In the production phase, the device safety engineering report is prepared by using the data collected during the phase. The report documents and highlights the device hazards. Finally, during the deployment phase, data concerning incidents, accidents, failures, etc., are collected, and safety professionals review any changes to the device. The device safety analysis is updated as necessary.

6.6 Software Issues in Cardiac Rhythm Management Products Safety

In regard to cardiac rhythm management products, there are many important issues to consider during their software safety analysis. These issues include marketing issues, technical issues, and management issues. It means that software developers must examine such issues with care in light of product context, environmental requirements, and constraints [14]. Two typical examples of the cardiac rhythm management systems or products are pacemakers and defibrillators used for providing electrical therapy to malfunctioning cardiac muscles. Nonetheless, four main elements of the marketing issues are shown in Figure 6.2 [3,14]. These include market requirements, regulatory requirements, product requirements, and legal requirements.

With respect to the market requirements, the sheer size of the cardiac rhythm management systems/products market is the sole important factor in their safety. For example, each year in the United States about half a million individuals experience a sudden cardiac death episode and around 31,000 individuals receive a defibrillator implant [3,14,15]. Furthermore, as per Refs. [3,14], the predictions for 1997 for the world

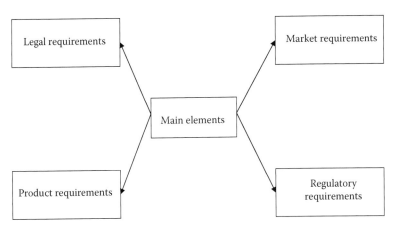

FIGURE 6.2
Main elements of the marketing issues.

market of cardiac rhythm management systems/products were around $3 billion.

The regulatory requirements are important as regulatory agencies, such as the FDA, demand that companies producing cardiac rhythm management systems have systematic and rigorous software development processes, including safety analysis. It is estimated that about half of the product/device development cycle is consumed by the regulatory acceptance process.

In regard to the product requirements, a typical cardiac rhythm management device or system is composed of electrical subsystems, mechanical subsystems, and advanced software that must function normally for its successful operation. Furthermore, as the software subsystem of modern medical devices is made up of around 500,000 lines of code, its safety, efficiency, and reliability to control both internal and external operations are very important.

Finally, legal requirements also are very important because of the life-or-death nature of cardiac rhythm management devices/systems, and the concerned regulations lead to highly sensitive legal requirements that involve manufacturers, patients and their families, regulatory bodies, etc. More specifically, in the event of injury to a patient or death, lawsuits by the regulatory bodies and the termination of products/devices by these bodies because of their safety-related problems are the driving force for the legal requirements.

The technical issues are an important factor as well during the software development process because the software maintenance and complexity-related concerns basically determine the analysis method and the incorporation process to be employed. With respect to maintenance, the code resulting from the required modifications of safety-critical software is one of the most critical elements. A study performed by one medical device software developer reported that out of the software-related change requests, and subsequent to the internal release of software during development, 41% were concerned with software safety [3,14]. In regard to complexity, modern medical devices can have a large number of parallel, asynchronous, and real-time software tasks reacting to randomly occurring external events. Thus, to ensure the timely and correct behavior of such complex software is quite difficult, as is the mitigation and identification of safety faults with correctness and timeliness.

Finally, the management issues are primarily concerned with making appropriate changes to the ongoing development process to include explicit safety analysis, thus requiring convincing justifications and a clear vision. In this case, it will certainly be helpful if the management is shown explicitly that the cost of performing software safety analysis during the development process can help to reduce the overall development cost, market losses, regulatory-related problems, and avoid potential legal costs.

6.7 Classifications of Medical Device Accident Causes and Legal Aspects of Medical Device Safety

There are many causes for the occurrence of medical device-related accidents. The professionals working in the area have classified these accident causes under the following seven classifications [16]:

1. Design defect
2. Manufacturing defect
3. Random component failure
4. Operator/patient error
5. Faulty calibration, preventive maintenance, or repair
6. Abnormal or idiosyncratic patient response
7. Malicious intent or sabotage

Additional information on the above classifications is available in Ref. [16].

One of the main objectives of system/product/device safety is to limit legal liability. Tort law complements safety regulations through the deterrent of manufacturing harmful medical devices, in addition to providing satisfactory compensation to injured individuals. The decision of the U.S. Supreme Court on the *Medtronic, Inc. v. Lohr* case, for Lohr, put additional pressure on companies to produce reliable and safe medical devices [17]. The case was filed by a Florida woman, Lora Lohr, who had a cardiac pacemaker implanted, which was produced by Medtronic, Inc., to regulate her abnormal heart rhythm. The pacemaker malfunctioned, and she alleged that the malfunction of the device was the result of defective design, manufacturing, and labeling [17]. The three commonly used theories to make manufacturers liable for injury caused by their products are shown in Figure 6.3 [3,17].

In the case of breach of warranty, it may be alleged under the following three scenarios [3,17]:

- Breach of an expressed warranty
- Breach of the implied warranty of merchantability
- Breach of the implied warranty of suitability for a specific use

For example, if a medical device caused injury to a person because of its inability to function as warranted, the manufacturer of that device faces liability under the breach of an expressed warranty scenario.

With respect to negligence, if the device manufacturer fails to exercise reasonable care or fails to meet a reasonable standard of care during the device handling, manufacturing, or distribution process, it could be liable for any

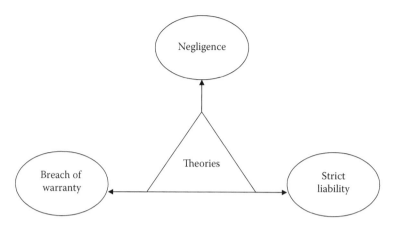

FIGURE 6.3
Commonly used theories to make manufacturers liable for injury caused by their products.

damages resulting from the device. Finally, in regard to strict liability, the basis for imposing it is that the manufacturer of a device is in the best position to reduce related risks.

6.8 Methods for Performing Medical System Safety Analysis and Considerations for Their Selection

There are many methods that can be used to perform safety analysis of medical systems/devices. Some of these methods include [1,3,13,18–21]:

- Operating hazard analysis
- Human error analysis
- Fault tree analysis
- Preliminary hazard analysis
- Failure modes and effect analysis
- Technic of operations review
- Interface safety analysis
- Root cause analysis

The first three of the above eight methods are presented below and the remaining five methods (i.e., preliminary hazard analysis, failure modes and effect analysis, technic of operations review, interface safety analysis, and root cause analysis) are described in Chapter 4.

6.8.1 Operating Hazard Analysis

This method focuses on hazards occurring from activities/tasks for operating system functions that happen as the system is used, transported, or stored. Normally, the operating hazard analysis (OHA) is started early in the system development cycle so that appropriate inputs to technical orders are provided, which in turn govern the testing of the system. The application of the OHA provides a basis for safety considerations, such as [3,13,18]:

- Design modifications for eradicating hazards
- Identification of item/system functions relating to hazardous occurrences
- Special safety procedures in regard to servicing, transporting, handling, storing, and training
- Development of emergency procedures, warning, or special instructions in regard to operation
- Safety devices and safety guards

It is to be noted that the analyst involved in the performance of OHA needs engineering descriptions of the system/device under consideration with available support facilities. Furthermore, OHA is carried out using a form that requires information on items such as the operational event description, hazard effects, hazard control, hazard description, and requirements.

Additional information on OHA is available in Refs. [13,18].

6.8.2 Human Error Analysis

This method is considered quite useful to identify hazards prior to their occurrence in the form of accidents. There could be two approaches to human error analysis: (1) Performing tasks to obtain first-hand information on hazards, and (2) observing workers during their work hours in regard to hazards. All in all, regardless of the performance of the human error analysis, it is strongly recommended to perform it in conjunction with failure modes and effect analysis and hazard and operability (HAZOP) analysis methods presented in Chapter 4.

Additional information on this method is available in Refs. [1,13,18–21].

6.8.3 Fault Tree Analysis

This is a widely used method to perform safety and reliability analysis of engineering systems in the industrial sector. The method was originally developed in the early 1960s to evaluate the safety of the Minuteman Launch

Control System [22]. Some of the main points concerned with fault tree analysis include [3,21]:

- It is an extremely useful analysis in the early design phases of new systems/devices/items.
- It allows users to evaluate alternatives as well as pass judgment on acceptable trade-offs among them.
- It can be used to evaluate certain operational functions (e.g., shutdown or start-up phases of facility/system/device operation).
- It is an effective tool to analyze operational systems/devices for desirable or undesirable occurrences.

Additional information on the method is available in Chapter 4 and in Ref. [22].

6.8.4 Considerations for the Selection of Safety Analysis Methods

Past experiences indicate that to perform an effective safety analysis of medical systems requires a careful consideration in the selection and implementation of appropriate safety analysis methods for given situations. Thus, questions, such as those listed below, should be asked with care prior to selection and implementation of safety analysis approaches for situations under consideration [3,18].

Who are the end results users?

What type of information, data, etc., are needed before the start of the study?

What mechanism is required to acquire information from subcontractors (if applicable)?

When are the results needed?

What is the exact time frame for the initiation of analysis and its completion, submission, review, and update?

Problems

1. Write an essay on medical systems safety.
2. List at least six facts and figures concerned with medical systems safety.
3. Discuss medical device hardware and software safety.
4. Discuss essential safety-related requirements for medical devices placed by government and other agencies.
5. Discuss safety in medical device life cycle.

6. What are the important software issues in cardiac rhythm management products safety?
7. Describe operating hazard analysis.
8. What are the classifications of medical device accident causes?
9. What are the commonly used theories to make manufacturers liable for injury caused by their products?
10. Discuss the following two classifications of medical device safety:
   a. Unconditional safety
   b. Descriptive safety

References

1. Goetsch, D. L., *Occupational Safety and Health*, Prentice Hall, Englewood Cliffs, NJ, 1996.
2. Ladou, J., Ed., *Introduction to Occupational Health and Safety*, The National Safety Council, Chicago, 1986.
3. Dhillon, B. S., *Medical Device Reliability and Associated Areas*, CRC Press, Boca Raton, FL, 2000.
4. Banta, H. D., The Regulation of Medical Devices, *Preventive Medicine*, Vol. 19, 1990, pp. 693–699.
5. *Medical Devices*, Hearings before the Subcommittee on Public Health and Environment, U.S. Congress Interstate and Foreign Commerce, Serial No. 93-61, U.S. Government Printing Office, Washington, D.C., 1973.
6. Schneider, P., Hines, M. L. A., *Classification of Medical Software*, Proceedings of the IEEE Symposium on Applied Computing, 1990, pp. 20–27.
7. Dhillon, B. S., *Reliability Technology in Health Care Systems*, Proceedings of the IASTED International Symposium on Computers, Advanced Technology in Medicine, and Health Care Bioengineering, 1990, pp. 84–87.
8. Casey, S., *Set Phasers on Stun: And Other True Tales of Design Technology and Human Error*, Aegean, Inc., Santa Barbara, CA, 1993.
9. *Mechanical Malfunctions and Inadequate Maintenance of Radiological Devices*, Medical Device Safety Report, prepared by the Emergency Care Research Institute, Plymouth Meeting, PA, 2001.
10. *Electric Beds Can Kill Children*, Medical Device Safety Report, prepared by the Emergency Care Research Institute, Plymouth Meeting, PA, 2001.
11. Fries, R. C., *Reliable Design of Medical Devices*, Marcel Dekker, New York, 1997.
12. Leitgeb, N., *Safety in Electromedical Technology*, Interpharm Press, Buffalo Grove, IL, 1996.
13. Roland, H. E., Moriarty, B., *System Safety Engineering and Management*, John Wiley & Sons, New York, 1983.
14. Mojdehbakhsh, R., Tsai, W. T., Kirani, S., Elliott, L., Retrofitting Software Safety in an Implantable Medical Device, *IEEE Software*, Vol. 11, January 1994, pp. 41–50.

15. Lowen, B., Cardiovascular Collapse and Sudden Cardiac Death, in *Heart Disease: A Textbook of Cardiovascular Medicine*, ed. E. Braunwald, W. B. Saunders, Philadelphia, 1984, pp. 778–803.

16. Brueley, M. E., *Ergonomics and Error: Who is Responsible?* Proceedings of the First Symposium on Human Factors in Medical Devices, 1989, pp. 6–10.

17. Bethune, J., Ed., On Product Liability: Stupidity and Waste Abounding, *Medical Device and Diagnostic Industry Magazine*, Vol. 18, No. 8, 1996, pp. 8–11.

18. System Safety Analytical Techniques, *Safety Engineering Bulletin*, No. 3, May 1971. Available from the Electronic Industries Association, Washington, D.C.

19. Gloss, D. S., Wardle, M. G., *Introduction to Safety Engineering*, John Wiley & Sons, New York, 1984.

20. Hammer, W., *Product Safety Management and Engineering*, Prentice Hall, Englewood, NJ, 1980.

21. Dhillon, B. S., *Advanced Design Concepts for Engineers*, Technomic Publishing, Lancaster, PA, 1998.

22. Dhillon, B. S., Singh, C., *Engineering Reliability: New Techniques and Applications*, John Wiley & Sons, New York, 1981.

7

Mining Equipment Safety

7.1 Introduction

Each year a vast sum of money is spent to produce various types of equipment for use in the mining sector around the globe. Today, the type of equipment used in the mining sector has come a long way since man first used tools made of flint and bone to extract ores from the Earth. Some examples of the type of equipment used in the mining sector are hoist controllers, crushers, dragline excavators, haul trucks, and mine carts.

Over the years, many mine accidents, to a certain degree, have occurred involving, directly or indirectly, mining equipment. In 1977, in order to improve safety in U.S. mines (including mining equipment safety), the U.S. Congress passed the Mine Safety and Health Act. As the result of this Act, the U.S. Department of Labor established an agency called Mine Safety and Health Administration (MSHA). The main goal of MSHA is to promote better health and safety conditions in the mining sector, reduce health-related hazards, and enforce compliance with mine safety and health standards [1,2].

This chapter presents various important aspects of mining equipment safety.

7.2 Facts and Figures

Some of the facts and figures, directly or indirectly, concerned with U.S. mining equipment safety include:

- During the period 1995 to 2005, 483 fatalities in the U.S. mining operations were concerned with equipment [1,2].
- In 2004, approximately 17% of the 37,445 injuries in the underground coal mines were connected to bolting machines [3].
- During the period 1990 to 1999, 197 equipment fires caused 76 injuries in the coal mining operations [4].

- A study conducted by the U.S. Bureau of Mines (now National Institute for Occupational Safety and Health (NIOSH)) revealed that equipment was the basic cause of injury in approximately 11% of mining accidents and a secondary causal factor in the occurrence of another 10% of the accidents [5–7].
- As per Ref. [8], during the period 1978 to 1988, maintenance activities in the mines accounted for approximately 34% of all lost-time injuries.
- As per Ref. [9], during the period 1983 to 1990, approximately 20% of the coal mine injuries occurred during the equipment maintenance activity or while using handheld tools.
- During the period 1990 to 1999, electricity was the fourth leading cause for the occurrence of fatalities in the mining industry [10].

7.3 Types of Mining Equipment Involved in Fatal Accidents and the Fatal Accidents' Breakdowns and Main Causes of Mining Equipment Accidents

Over the years, many fatal accidents involving various types of mining equipment have occurred. For example, a MSHA study reported that, during the period 1995 to 2005, there were 483 equipment-related fatal accidents in U.S. mining operations [11]. The type of equipment involved is listed below and its corresponding fatal accidents, for the specified period, are in parentheses [2,11]:

- Haul truck (108)
- Conveyor (45)
- Front-end loader (41)
- Miner (30)
- Dozer (28)
- Drill (16)
- Shuttle car (13)
- Roof bolter (7)
- LHD (load-haul-dump) (6)
- Forklift (5)
- Longwall (5)
- Hoisting (2)
- Miscellaneous equipment (177)

Furthermore, the percentage distributions of the above accidents in regard to the type of equipment involved (in parentheses) were: 22.36% (haul truck), 9.32% (conveyor), 8.49% (front-end loader), 6.21% (miner), 5.8% (dozer), 3.31% (drill), 2.69% (shuttle car), 1.45% (roof bolter), 1.24% (LHD), 1.04% (fork-lift), 1.04% (longwall), 0.41% (hoisting), 36.65% (misc. equipment) [11]. It means over 50% of the fatal accidents were due to five types of equipment only (i.e., haul truck, conveyor, front-end loader, miner, and dozer). Over the years, many studies have been performed to highlight main causes of mining equipment accidents. One such study was performed by the U.S. Bureau of Mines (now NIOSH). The study identified seven main causes for the occurrence of mining equipment accidents: restricted visibility, poor original design or redesign, exposed sharp surfaces/pinch points, poor control display layout, hot surfaces/exposed wiring, poor ingress/egress design, and unguarded moving parts [5].

7.4 Mining Ascending Elevator Accidents, and Fatalities and Injuries Due to Drill Rig, Haul Truck, and Crane Contact with High Tension Power Lines

Over the years many ascending elevator accidents in mining operations have occurred and a number of them have resulted in serious injuries or fatalities. These accidents have occurred on counterweighted elevators due to electrical, mechanical, and structural failures. Past experiences indicate that, although the elevator cars have safe ties that grip the guide rails and stop a falling car, such devices fail to provide an appropriate level of protection in the upward direction.

In 1987, an ascending elevator car accident at a Pennsylvania coal mine caused extensive structural damage and disabled the elevator for many months [12]. As the result of this accident, in order to provide effective ascending car over speed protection for current and new mine installations, the Pennsylvania Bureau of Deep Mine Safety established an advisory committee to evaluate devices on the market. After an extensive investigation, the committee recommended four protective methods (Figure 7.1), in addition to requirements that all new elevators must have a governor rope monitoring device, back out of over travel switch, and a manual reset [2,12].

Additional information on the methods shown in Figure 7.1 is available in Refs. [12–14].

Past experience indicates that overhead or high-tension power lines present a serious electrocution hazard to people employed in a variety of industrial sectors because equipment such as haul trucks, drill rigs, and cranes are frequently exposed to these power lines. When contacting with the power lines, this type of equipment becomes susceptible to a high voltage, and simultaneous contact to the "hot" frame and ground by people can lead to dangerous burns and electric shocks.

The industrial sectors where the risk of occurrence of such accidents is greatest include mining, agriculture, and construction.

In the United States, around 2300 accidental overhead power line contacts occur each year [15]. The U.S. mining industry reported at least 94 mobile equipment overhead line contact accidents during the period 1980 to 1997 [15].

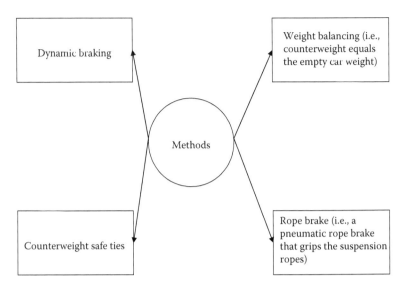

FIGURE 7.1
Mining ascending elevator protective methods.

These accidents resulted in 114 injuries and about 33% of them were fatal. Most of these accidents involved drills (14%), dump bed trucks (24%), and cranes (47%) [15].

7.5 Programmable Electronic-Related Mining Mishaps and Lessons Learned

Nowadays, programmable electronic-related mishaps have become an important issue in the mining industry. The serious thinking of the MSHA about the functional safety of programmable electronics (PE)-based mining systems started in 1990 because of an unplanned longwall shield mishap [16]. During the period 1995 to 2001, there were 11 PE-related mining incidents in the United States; four of them caused fatalities [17,18]. During the same period (i.e., 1995–2001), there were 71 PE-related incidents in underground coal mines in New South Wales, Australia [17]. A study of both these data sets reported that most of incidents or mishaps involved sudden startups or movements of PE-based mining systems.

In 1991, MSHA conducted a study of longwall installations in regard to programmable electronics. The study reported that about 35% had experienced sudden movements basically due to four problems: operator errors, software programming errors, water ingress, and defective or sticking solenoid valves. A detailed analysis of the U.S. and Australian data sets revealed that there were four factors (i.e., improper operation, software, solenoid valves, and water ingress) that contributed to PE-based mishaps [16,19]. However, the solenoid valves-related problems were the main contributing factor to PE-based incidents or mishaps.

In response to the longwall-related mishaps, MSHA recommended improvements in four areas: operator training, maintaining integrity of enclosure sealing, timely maintenance, and maintaining alertness for abnormal operational sequences that might be indicative of a software problem.

Subsequently, MSHA proposed a safety framework largely based on the International Electrotechnical Commission (IEC) 61508 safety life cycle [20], in order to overcome the shortcomings of the above approach.

Over the years, many valuable lessons have been learned to address programmable electronic mining system safety-related issues. Most of these lessons include [17]:

- Establishing appropriate terminology, concepts, and definitions as early as possible
- Involving all appropriate elements of the industrial sector early and on a continuous basis
- Holding industry workshops as considered appropriate
- Identifying and clearly understanding all related issues and perceptions
- Making use of appropriate scenarios to convey information
- Decomposing the problem into manageable parts
- Clearly separating the associated concerns

7.6 Equipment Fire-Related Mining Accidents and Mining Equipment Fire Ignition Sources

Over the years, there have been many equipment fire-related mining accidents that have caused injuries. For example, during the period 1990 to 1999 in U.S. coal mines there were 197 equipment fires that caused 76 injuries [4]. For further study, NIOSH grouped these fires under the following four categories [4]:

Category I: Surface equipment fires at underground coal mines. There were 140 equipment fires and 56 of them caused 56 injuries.

Category II: Underground equipment fires. There were 26 equipment fires and 10 of them caused 10 injuries.

Category III: Prep plant fires. There were 17 equipment fires and 6 of them caused 6 injuries.

Category IV: Surface coal mine equipment fires. There were 14 equipment fires and 4 of them caused 4 injuries.

Over the years, many studies have been performed to identify ignition sources for mining equipment fires. The four main ones are [4]:

1. Hydraulic fluid/fuel on equipment hot surfaces
2. Flame cutting/welding spark/slag
3. Electric short/arcing
4. Engine malfunction

Additional information on the above ignition sources is available in Ref. [4].

7.7 Strategies to Reduce Mining Equipment Fires and Useful Guidelines to Improve Electrical Safety in Mines

Over the years, many different ways have been explored to reduce mining equipment fires and injuries. Some of the better strategies/methods are [4]:

- Provide frequent and necessary emergency preparedness training to all equipment operators.
- Perform equipment hydraulic, fuel, and electrical system inspections thoroughly and frequently.
- Develop effective equipment/cab fire detection systems having an audible/visible cab alarm.
- Aim to develop new technologies for fire barriers, emergency hydraulic line drainage/safeguard system, and emergency engine/ pump shutoff.
- Improve as necessary equipment/cab fire suppression/prevention systems.

Over the years many guidelines to improve general electrical safety in the mining industry (including equipment) have been proposed. Five of these guidelines considered most useful include [10,21]:

- Make necessary improvements in electrical maintenance schedules and procedures.
- Make use of appropriate power line avoidance devices as much as possible.

- Target appropriate training in problem areas.
- Make improvements in equipment/system design in general and in regard to electrical safety in particular.
- Provide appropriate power line awareness training to concerned individuals.

7.8 Human Factors-Related Design Tips for Safer Mining Equipment

A number of studies conducted over the years have indicated that equipment is the primary cause of injury in around 11% of mining accidents and a secondary cause in another 10%. Therefore, it is of utmost importance that new mining equipment clearly incorporate good human-factors design criteria that maximize the safety of all mine workers. In this regard, human factors-related tips considered most useful are divided into four areas as shown in Figure 7.2 [6,21].

Five tips concerning the control design area include [6,21]:

1. Ensure that all types of design controls clearly comply with anthropometric data on human operators.
2. Ensure that all involved operators can identify the appropriate controls, accurately and quickly.

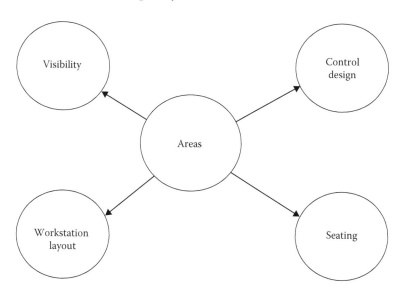

FIGURE 7.2
Areas of human factors-related tips for safer mining equipment.

3. Ensure that a vehicle's or a part's speed is proportional to the control displacement from its rest position and in the same direction.
4. Ensure that controls have adequate resistance to decrease the possibility of inadvertent activation by the weight of a foot or hand.
5. Ensure that design controls can clearly withstand or guard against abuse, such as from the forces imposed during a panic response in an emergency situation or from falling ribs and roof.

Five tips associated with the seating area include [21,22]:

1. Design the seat in such a way that mine workers at large can easily replace or maintain it.
2. Ensure that the seat does not hinder the ability of the involved operator to control the machine/equipment.
3. Ensure that the seat provides essential design features to guard against shocks due to rough roads and minor collisions that tend to unseat the person occupying the seat.
4. Ensure that the seat does not hinder the operator's ability to exit or enter the workstation.
5. Ensure that the seat properly adjusts and fits to body dimensions, distributes weight to relieve pressure points, and supports posture.

Four tips concerning the workstation area include [6,21]:

1. Anticipate all types of potential safety-related hazards and necessary emergency measures before starting the design process.
2. Ensure that the workstation fits operators effectively from the 5th to 95th percentile range.
3. Aim to distribute workload as evenly as possible between feet and hands.
4. Ensure that the relative placement of displays and controls for similar equipment is maintained effectively.

Finally, two tips concerning the visibility area are [6,21]:

1. Ensure that there is sufficient contrast between the luminance of the object or location of interest and the surrounding background so that the required task can be performed effectively and safely.
2. Ensure that the workstation provides an unobstructed line of sight to objects/locations that should be clearly visible to perform a task effectively and safely.

7.9 Hazardous Area Signaling and Ranging Device (HASARD) Proximity Warning System

Over the years, many surface and underground mining workers working close to machinery and powered haulage have been killed or permanently disabled. For example, as per Ref. [22], each year in surface mining operations about 13 persons are killed by being run over or pinned by mobile equipment. Furthermore, in underground mining operations in the United States, during the period 1988 to 2000, 23 fatalities were associated with mining workers getting crushed, caught, or pinned by continuous mining equipment/systems.

A subsequent analysis of these fatalities revealed that, time to time, mining workers become totally preoccupied with operating their own equipment and fail to realize when they stray into or are subjected to hazardous conditions.

In order to overcome problems such as these, NIOSH developed an active proximity warning system known as hazardous area signaling and ranging device (HASARD). Over the years, this system has repeatedly proved to be a very useful tool to warn mine workers when they approach hazardous areas around heavy mining equipment and other hazardous areas. The system is composed of the following two subsystems [21,22]:

Transmitter: It generates a 60 kHz magnetic field with the aid of one or more wire loop antennas. In turn, each antenna is adjusted to establish a magnetic field pattern for each hazardous area, as the need arises.

Receiver: This is a magnetic field meter and is worn by the mining workers. It compares the received signal with preset levels, which are calibrated to highlight dangerous levels. Furthermore, the receiver outputs can be made to stop ongoing equipment operations and can include visual, audible, and vibratory indicators.

Additional information on HASARD proximity warning system is available in Ref. [22].

7.10 Useful Methods to Perform Mining Equipment Safety Analysis

There are a large number of methods available in the published literature that can be used to conduct mining equipment safety analysis [23]. Nine of these methods considered most useful to perform mining equipment safety analysis are (1) management oversight and risk tree (MORT) analysis, (2) consequence analysis, (3) binary matrices, (4) human reliability analysis (HRA), (5) preliminary hazards analysis, (6) failure modes and effect analysis (FMEA), (7) root cause analysis, (8) technic of operations review, and (9) hazards and operability analysis. The first four methods are presented below and the remaining five methods are described in Chapter 4.

7.10.1 Management Oversight and Risk Tree Analysis

This is a comprehensive safety assessment method that can be applied to any mine safety program and is based on a document prepared by W. G. Johnson (director at the Atomic Energy Commission) in 1973 [24]. The method focuses on administrative or programmatic control of hazardous conditions and is particularly designed to highlight, evaluate, and prevent the occurrence of safety errors, omissions, and oversights by workers and management that can lead to accidents.

The following nine steps are followed to perform management oversight and risk tree (MORT) analysis [25]:

Step 1: Obtain sufficient working knowledge of the system/equipment under study.

Step 2: Select the accident for analysis.

Step 3: Identify potential hazardous energy flows and barriers related to the accident sequence.

Step 4: Document necessary information in the standard MORT-type analytical tree format.

Step 5: Determine all possible factors that can cause initial unwanted energy flow.

Step 6: Document the safety program elements that are considered to be less than sufficient in regard to the unwanted energy flow.

Step 7: Continue conducting analysis of the safety program elements with respect to the rest of the unwanted energy flows (if any).

Step 8: Determine the management system factors associated with the potential accident.

Step 9: Review the accomplished analysis for all safety program elements that could reduce the likelihood of the occurrence of the potential accident.

Some of the advantages and disadvantages of the MORT analysis method are listed below [24].

Advantages

- An effective and comprehensive approach that attempts to review each and every aspect of safety in any type of work.
- Results of MORT analysis can suggest appropriate improvements to an existing safety program that could be quite helpful in decreasing injuries, reducing property damage, and saving lives.
- Useful to evaluate all three aspects of an industrial system (i.e., hardware, human, and management) as they collectively cause accidents.

Disadvantages

- Creates a vast amount of complex detail.
- Emphasizes management's responsibility to provide a safe work environment.
- A time-consuming method.

Information on the application of the MORT analysis to a mining system is available in Ref. [25].

7.10.2 Binary Matrices

This is a useful, logical, and qualitative method to identify system interactions [26]. The method can be used during the safety analysis system-description stage, or as a final checkpoint in a preliminary hazards analysis or FMEA to ensure that all important system dependencies have been considered in the analysis effectively.

The specific tool employed in binary matrices is the binary matrix that contains information on the relationships between system elements. The primary objective of binary matrix is to highlight the one-on-one dependencies that exist between system elements. All in all, this matrix serves merely as a useful tool to "remind" the analyst that failures in one part of a system/equipment may effect the normal operation of other subsystems in completely distinct areas.

Information on the application of the binary matrices method to a mining system is available in Ref. [25].

7.10.3 Human Reliability Analysis (HRA)

HRA may simply be described as the study of human performance within the framework of a complex man–machine operating system. The method is considered quite useful in developing qualitative information concerning the causes and effects of human errors in specific situations. HRA is conducted by following the steps shown in Figure 7.3 [25].

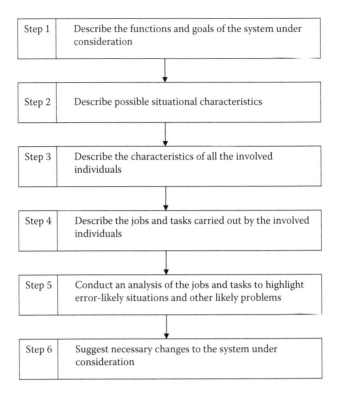

FIGURE 7.3
Human reliability analysis steps.

Additional information on the steps shown in Figure 7.3 is available in Refs. [25,27]. The application of the method to a mining problem is demonstrated in Ref. [25].

7.10.4 Consequence Analysis

This method is concerned with determining the impact of an undesired event on items such as adjacent property, the environment, or people. Four typical examples of an undesired event are explosion, fire, the release of toxic material, and projection of debris. Some of the primary consequences of concern in the area of mining are fatalities, injuries, and losses due to property/equipment damage and operational downtime.

All in all, consequence analysis is one of the intermediate steps of safety analysis as accident consequences are generally determined, initially using methods such as failure modes and effect analysis or preliminary hazards analysis. Additional information on consequence analysis is available in Ref. [27] and information on its application in the area of mining is given in Ref. [25].

Problems

1. List at least six important facts and figures, directly or indirectly, concerning mining equipment safety.
2. List at least 10 types of mining equipment involved in fatal accidents.
3. What are the main causes of mining equipment accidents?
4. Discuss mining ascending elevator accidents.
5. What are the seven important lessons learned in addressing programmable electronic mining system safety-related issues?
6. What are the main mining equipment fire ignition sources?
7. What are the important strategies to reduce mining equipment fires?
8. Discuss human factors-related design tips for safer mining equipment.
9. Describe hazardous area signaling and ranging device (HASARD) proximity warning system.
10. List at least seven methods that can be used to perform mining equipment safety analysis and then describe management oversight and risk tree (MORT) analysis.

References

1. Mine Safety and Health Administration (MSHA), U.S. Department of Labor, Washington, D.C. Available online at http:www.msha.gov/.
2. Dhillon, B. S., *Mine Safety: A Modern Approach*, Springer-Verlag, London, 2010.
3. Burgess-Limerick, R., Steiner, L., Preventing Injuries: Analysis of Injuries Highlights High Priority Hazards Associated with Underground Coal Mining Equipment, *American Longwall Magazine*, August 2006, pp. 19–20.
4. De Rosa, M., Equipment Fires Cause Injuries: Recent NIOSH Study Reveals Trends for Equipment Fires at U.S. Coal Mines, *Coal Age*, No. 10, 2004, pp. 28–31.
5. Saunders, M. S., Shaw, B. E., *Research to Determine the Contribution of System Factors in the Occurrence of Underground Injury Accidents*, Report No. USBM OFR 26-89, U.S. Bureau of Mines (USBM), Washington, D.C., 1988.
6. Unger, R. L., Tips for Safer Mining Equipment, U.S. Department of Energy's Mining Health and Safety Update, Vol. 1, No. 2, 1996, pp. 14–15.
7. *What Causes Equipment Accidents?* National Institute for Occupational Safety and Health (NIOSH), 2008. Available online at http://www.cdc.gov/niosh/mining/topics/machinesafety/equipmentdsgn/equipment accident.
8. MSHA Data for 1978–1988, Mine Safety and Health Administration (MSHA), U.S. Department of Labor, Washington, D.C.
9. Rethi, L. L., Barett, E. A., *A Summary of Injury Data for Independent Contractor Employees in the Mining Industry from 1983–1990*, Report No. USBMIC 9344, U.S. Bureau of Mines, Washington, D.C., 1983.

10. Cawley, J. C., Electrical Accidents in the Mining Industry: 1990–1999, *IEEE Transactions on Industrial Application*, Vol. 39, No. 6, 2003, pp. 1570–1576.

11. Kecojevic, V., Komljenovic, D., Groves, W., Rodomsky, M., An Analysis of Equipment-Related Fatal Accidents in U.S. Mining Operations: 1995–2005, *Safety Science*, Vol. 45, 2007, pp. 864–874.

12. Barkand, T. D., Ascending Elevator Accidents: Give the Miner a Brake, *IEEE Transactions on Industry Applications*, Vol. 28, No. 3, 1992, pp. 720–729.

13. Nederbragt, J. A., Rope Brake: As Precaution Against Overspeed, *Elevator World*, Vol. 7, 1989, pp. 6–7.

14. Barkand, T. D., Helfrich, W. J., Application of Dynamic Braking to Mine Hoisting Systems, *IEEE Transactions on Industry Applications*, Vol. 24, No. 5, 1988, pp. 507–514.

15. Sacks, H. K., Cawley, J. C., Homce, G. T., Yenchek, M. R., Feasibility Study to Reduce Injuries and Fatalities Caused by Contact of Cranes, Drill Rigs, and Haul Trucks with High-Tension Lines, *IEEE Transactions on Industry Applications*, Vol. 37, No. 3, 2001, pp. 914–919.

16. Dransite, G. D., *Ghosting of Electro-Hydraulic Long Wall Shield Advanced Systems*, Proceedings of the 11th West Virginia University International Electro Technology Conference, 1992, pp. 77–78.

17. Sammarco, J. J., *Addressing the Safety of Programmable Electronic Mining Systems: Lessons Learned*, Proceedings of the 37th IEEE Industry Applications Society Meeting, 2003, pp. 692–698.

18. *Fatal Alert Bulletins, Fatal Grams and Fatal Investigation Reports*, Mine Safety and Health Administration (MSHA), Washington, D.C., May 2001. Available online at http://www.msha.gov/fatals/fab.htm.

19. Waudby, J. F., *Underground Coal Mining Remote Control of Mining Equipment: Known Incidents of Unplanned Operation in New South Wales (NSW) Underground Coal Mines*, Dept. of Mineral Resources, NSW Department of Primary Industries, Maitland, NSW, Australia, 2001.

20. IEC 61508, Parts 1-7, *Functional Safety of Electrical/Electronic/Programmable Electronic Safety-Related Systems*, International Electrotechnical Commission (IEC), Geneva, Switzerland, 1998.

21. Dhillon, B. S., *Mining Equipment Reliability, Maintainability, and Safety*, Springer-Verlag, London, 2008.

22. *Hazardous Area Signalling and Ranging Device (HASARD)*, Pittsburgh Research Laboratory, National Institute for Occupational Safety and Health (NIOSH), Atlanta, GA.

23. Dhillon, B. S., *Engineering Safety: Fundamentals, Techniques, and Applications*, World Scientific Publishing, River Edge, NJ, 2003.

24. Johnson, W. G., *The Management Oversight and Risk Tree-MORT*, Report No. SAN 821-2, U.S. Atomic Energy Commission, Washington, D.C., 1973.

25. Daling, P. M., Geffen, C. A., *User's Manual of Safety Assessment Methods for Mine Safety Officials*, Report No. BuMines OFR 195(2)-83, U.S. Bureau of Mines, Department of the Interior, Washington, D.C., 1983.

26. Cybulskis, P., et. al, *Review of Systems Interaction Methodologies*, Report No. NUREG/CR-1896, Battelle Columbus Laboratories, Columbus, OH, 1981.

27. *Guidelines for Consequences Analysis of Chemical Releases*, American Institute of Chemical Engineers, New York, 1999.

8

Robot and Software Safety

8.1 Introduction

Nowadays, robots are increasingly being used in the industrial sector to perform various types of tasks including spot welding, arc welding, materials handling, and routing. Currently, there are over 1 million robots in use in the industrial sector throughout the world. The history of robot safety may be traced back to the 1980s with the development of the American National Standard for Industrial Robots and Robot Systems: Safety Requirements [1], and the Japanese Industrial Safety and Health Association document entitled "An Interpretation of the Technical Guidance on Safety Standards in the Use, etc., of Industrial Robots" [2].

Nowadays, information on robot safety is widely available in the form of journal articles, book chapters, conference proceedings articles, technical reports, etc., as robot safety has become a pressing issue [3–5].

Today, computers have become an important element of day-to-day life and they are made up of both hardware and software components. Nowadays, much more money is spent to develop computer software than hardware. Needless to say, software has become a driving force in the computer industrial sector along with growing concerns for its safe functioning. In many applications, its proper functioning is so important that a simple malfunction or failure can lead to a large-scale loss of lives. For example, Paris commuter trains serving about 1 million passengers daily very much depend on software signaling [6].

Over the years, professionals working in the area of robotics and software have developed various types of methods and approaches for ensuring robot and software safety. This chapter presents various important aspects of robot and software safety.

8.2 Robot Safety-Related Facts, Figures, and Examples

Some of the facts, figures, and examples directly or indirectly concerned with robot safety include:

- The first robot-related fatal accident occurred in Japan in 1978 and in the United States in 1984 [7,8].
- During the period 1978 to 1987, there were 10 robot-related fatal accidents in Japan [7].
- In 1987, a study of 32 robot accidents that occurred in the United States, Japan, Sweden, and West Germany reported that line workers were at the greatest risk of injury followed by workers concerned with the maintenance activity [8].
- As per Ref. [9], a carousel operator was killed when the person's foot accidentally tripped a light sensor that, in turn, caused a computer-controlled robotic platform to fall onto the top of the operator's head.
- A material-handling robot was functioning in its automatic mode and a worker violated all involved safety devices to enter its work cell. The worker got trapped between the robot and a post anchored to the ground and got injured and died a few days later [5,7–15].
- A robot serving the line stopped momentarily on a program point and a worker climbed onto a conveyor belt in full motion to recover a faulty component. When the robot recommenced movement, the worker was crushed to death by the robot [5,7–15].
- A study revealed that 12 to 17% of the accidents in the industrial sector using advanced manufacturing technology were associated with automated production equipment [14,15].
- A worker stepped between a robot and the machine (a planer) it was servicing, and switched off the power to the circuit transmitting/activating signal from the machine (planer) to the robot. After completing the required tasks, the worker turned on the power to the same circuit while still being in the work zone of the robot. When the robot recommenced its operation, it crushed the worker in question to death against the planer [5,7–15].
- A worker switched on a welding robot while another person was still in the work zone of the robot. The robot pushed that individual into the positioning fixture and the individual subsequently died of injuries [5,7–15].
- A maintenance worker climbed over a safety fence without switching off power to the robot and carried out various tasks in the work zone of the robot while it was temporarily stopped. When the robot recommenced its operation, it pushed the maintenance

worker into a grinding machine and the worker subsequently died of injuries [5,7–15].

8.3 Robot Accident Classifications and Causes of Robot Hazards

There are many different types of robot accidents. They can be grouped under the following four broad classifications [16]:

Trapping/crushing accidents. These accidents are concerned with situations where a person's limb or other body part is trapped between the arm of a robot and other peripheral equipment or the person is driven physically into and crushed by other peripheral equipment.

Collision/impact accidents. These accidents are concerned with situations where unpredicted part failures, movements, or unpredicted program changes related to the robot arm or peripheral equipment result in contact accidents.

Mechanical part accidents. These accidents are concerned with situations where the breakdown of the robot's parts, power source, peripheral equipment, or its tooling or end-effectors occurs. Some typical examples of mechanical failures are the failure of gripper mechanism, failure of end-effect or power tools, and release of components or parts.

Miscellaneous accidents. These include all those accidents that cannot be included into the above three classifications.

There are many causes of robot hazards. The main ones are shown in Figure 8.1 [16,17]. Control errors are an important cause of robot hazards and they can occur due to various reasons including faults in the hydraulic, pneumatic, or electrical controls associated with the robot system.

Human errors are another important cause of robot hazards and a common human error associated with robots is the incorrect activation of the "teach pendant" or control panel. Some other examples of human errors are interfacing activated peripheral equipment, placing oneself in a hazardous position while programming the robot, and connecting live input–output sensors to the microprocessor or a peripheral. Mechanical failures also can result in robot hazards because functioning programs may not account for cumulative mechanical part/component failure, thus making it possible for the occurrence of unexpected or faulty operation.

Environmental sources are also an important cause of robot hazards. For example, radio frequency or electromagnetic interference (transient signals) can exert an undesirable influence on robot operation and increase the

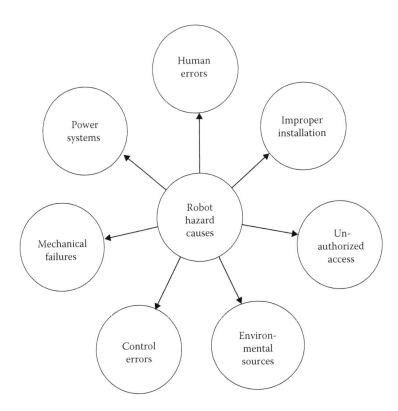

FIGURE 8.1
Main causes of robot hazards.

probability of injury to people working in the area. The unauthorized entry into the safeguarded robot zone is hazardous because the person concerned may be unfamiliar with the activation status or the safeguards in place.

Improper installation can cause various types of robot hazards as well. More specifically, the equipment layout, requirement design, utilities, and facilities of a robot system, if executed incorrectly, can lead to inherent hazards. Finally, power systems are another important cause of robot hazards because hydraulic, pneumatic, or electrical power sources that have malfunctioning transmission or control components/parts in the robot power system can cause a disruption in electrical signals to power supply/control lines.

8.4 Safety Considerations in Robot Life Cycle

Robot life cycle may be divided into four phases, as shown in Figure 8.2 [10]. Each of these phases is described below [2,5,18–20].

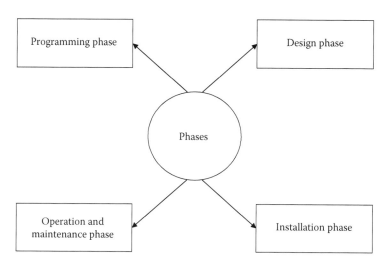

FIGURE 8.2
Robot life cycle phases.

8.4.1 Design Phase

In this phase, the safety considerations may be grouped under the following three classifications:

Classification I: Software. This classification includes safety considerations, such as having a standby power source for robots operating with programs in random access memory, prohibiting a restart by merely resetting a single switch, having built-in safety commands, using a procedure or checks to determine why a failure occurred, examining periodically the built-in self-checking software for safety, and providing a robot motion simulator.

Classification II: Electrical. This classification includes safety considerations, such as designing wire circuitry capable of stopping the robot's movement and locking its brakes, minimizing the effects of electromagnetic and radio frequency interferences, having built-in hose and cable routes using adequate insulation, sectionalization, and panel covers, and having a fuse "blow" long before human crushing pressure is experienced.

Classification III: Mechanical. This classification includes safety considerations, such as having mechanisms for releasing the stopped energy; designing teach pendant ergonomically; putting guard belts on items, such as gears, belts, and pulleys; having several emergency stop buttons; and eliminating sharp corners.

8.4.2 Installation Phase

There are many installation phase robot safety considerations. Some of these include installing the necessary interlocks to interrupt robot motion, placing robot controls outside the hazard zone, using vibration-reducing pads when appropriate, installing interlocks and sensing devices, distancing circuit boards from electromagnetic fields, installing electrical cables according to electrical codes, ensuring the visibility and accessibility of emergency stops, placing an appropriate shield between the robot and personnel, providing adequate illumination to personnel concerned with robots, and identifying the danger zones with the aid of signs, line markings, codes, etc.

8.4.3 Programming Phase

There are also many programming phase safety considerations for robots. Some of these are marking the programming positions, pressure sensitive mats on the floor at the position of the programmer, mandatory reduced speed, hold-to-run buttons, adjustable limits for robot axes, and a manual programming device containing an emergency off switch.

8.4.4 Operation and Maintenance Phase

This is a very important phase of the robot life cycle. Many robot hazards are confined to this phase. Some of the safety measures associated with this phase include:

- Ensuring the operational readiness of all associated safety devices.
- Ensuring the functionality of all emergency stops.
- Investigating any fault or unusual robot motions promptly.
- Blocking out all associated power sources during the maintenance activity.
- Performing preventive maintenance regularly and using only the approved spare parts.
- Developing appropriate safety operations and maintenance procedures.
- Minimizing the potential energy of an unexpected motion by having the robot arm extended to its maximum reach position.
- Minimizing the risk of fire by using nonflammable liquids for lubrication and hydraulics.
- Ensuring that only authorized and trained personnel operate and maintain robots.
- Providing appropriate protective gear to all involved personnel.

- Posting the operating weight capacity of the robot.
- Keeping (if possible) at least one extra person in the vicinity during the robot repair process.

8.5 Human Factors Issues in Robotic Safety

One of the most important factors in robotic safety is the issue of human factors and it has to be considered with care. This issue is discussed under the following areas [21,22]:

Human–robot interface design. In this case, the primary objective is to develop human–robot interface design in such a way that the occurrence of human error is at minimum. Some of the steps that can help to prevent the occurrence of human error are analyzing human actions during robotic processes, paying proper attention to all types of layouts, designing hardware and software with the intention to reduce the occurrence of human errors, and considering factors such as weight, hand-held device's shape and size, layout of buttons, flexibility of connectors, and the readability of buttons' functional descriptions.

Document preparation. This is concerned with the quality of documentation for potential robot users. These documents must be developed by considering the qualifications and experience of potential robot users, such as operators, programmers, and maintainers. More specifically, during the preparation process of such documents factors, such as easily understandable information, completeness of information, inclusion of pictorial descriptions at appropriate places, and inclusion of practical exercises, must be considered with care.

Methods analysis. In this case, the past experiences indicate that the classical industrial engineering approach to methods analysis is quite effective to improve robot safety in regard to human factors. Two important examples are flow process charts and multiple-activity process charts.

Future considerations. In this case, two of the important future considerations in improving robot safety in regard to human factors are considering human factors in factory design with respect to possible robot applications and the application of artificial intelligence to the worker–robot interface.

Miscellaneous considerations. In this case, the miscellaneous considerations include the preproduction testing of robotic systems; examination of factors, such as temperature, lighting, and noise; and the analysis of ambient conditions and layout.

8.6 Robot Safeguard Methods and a
Methodology for Safer Robot Design

There are many robot safeguard methods. Six of these methods are intelligent systems, warning signs, flashing lights, infrared light array, electronic devices, and physical barriers [5,23]. The intelligent systems use intelligent control systems for safeguarding. They use avenues, such as hardware, software, and sensing, in making decisions.

The warning signs are used in environments where robots, by virtue of their speed, inability, and size to impart excessive force, cannot injure humans. Nonetheless, past experiences indicate that the warning signs are extremely useful for all types of robot environments. The flashing lights are installed at the perimeter of the robot-working zone or on the robot itself to alert people that robot programmed motion is happening or could happen any moment.

In the case of infrared light arrays, the commonly used linear arrays of infrared sources are light curtains. Although the past experiences indicate that the light curtains are quite reliable, from time to time false triggering may occur due to factors such as smoke, heavy dust, or flashing lights because of misalignment of system parts. Electronic devices are basically the application of ultrasonic for perimeter control to have protection from intrusion. The perimeter control electronic barriers use active sensors to detect intrusions.

Finally, the basic objective of physical barriers is to stop humans from reaching over, under, around, or through the barrier into the forbidden work zone of a robot. Some examples of physical barriers are chain link fences, safety rails, plastic safety chains, and tagged rope barriers.

Over the years, various methodologies have been developed for safer robot design. One of these methodologies is composed of four basic steps, as shown in Figure 8.3 [24].

The first step, Determine task and safety requirements, is concerned with determining all appropriate task and safety-related requirements. The task requirements relate to attributes such as strength, agility, dexterity, robot position and force resolution, and allowable robot weight. The first four of these attributes depend on factors: expected payload, robot endrin speed, workspace size and types of robot motion, and sensor accuracy, respectively. The determination of safety requirements involves the specifications of the values of metrics, such as [24]:

- Maximum value of force during an impact with a robot element.
- Maximum value of force due to the weight of free element of robot.
- Maximum value of torque due to the weight of the free element of robot.

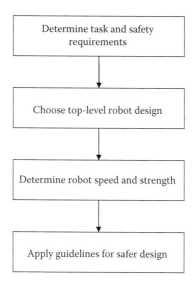

FIGURE 8.3
A four-step methodology for safer robot design.

- Maximum value of contact stress a robot can generate statically.
- Maximum value of force during an impact with a projectile released by a robot.
- Maximum value of contact stress during an impact with a robot element.
- Maximum value of contact stress during an impact with a projectile let go by a robot.

It simply means that the designer must establish the most appropriate overall safety level for the robot in regard to metrics such as listed above.

The second step, Choose top-level robot design, is basically concerned with choosing the type of robot and the basic parameters characterizing its geometry (e.g., link lengths and joint motion range). The third step, Determine robot speed and strength, is concerned with determining robot speed and strength that effectively ensure the task will be accomplished while satisfying given safety-related requirements. The execution of this step can be divided into the following four areas [24]:

1. Developing robot force-velocity-energy model
2. Plotting the safety and task envelopes on the safety diagram
3. Selecting the appropriate actuator and transmission characteristics to meet target specifications
4. Identifying the proper target pay load and velocity specifications

Finally, the fourth step, Apply guidelines for safer design, is concerned with applying the appropriate guidelines for safer design. These guidelines include actions, such as applying maximum padding, maximizing robot accuracy and dexterity, adhering to standard machine safety practices, eliminating all pinch points and maximizing vise radii, ensuring adequate contact area, and minimizing robot mass and inertia.

8.7 General Guidelines to Reduce Robot Safety Problems

Over the years various organizations and professionals have developed general guidelines to reduce robot safety-related problems. In particular, the National Institute for Occupational Safety and Health (NIOSH) has developed a set of general guidelines, divided into three areas, as shown in Figure 8.4, to reduce the risk of robot accidents [25].

The area, Robotic system design, contains guidelines, such as [10,25]:

- Providing effective clearance distances around the moving parts of robotic systems.
- Providing effective illumination in control and operational areas of robotic systems so that items, such as levers, written instructions, and buttons, are clearly visible.
- Incorporating an appropriate backup to items, such as motion sensors, floor sensors, electrical interlocks, or light curtains that stops the robot instantly whenever the barrier is crossed by a person.
- Including barriers between any free-standing objects and robotic equipment so that humans cannot get between any robot part and the "pinch points."

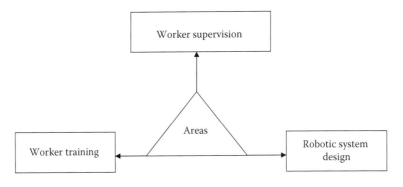

FIGURE 8.4
The National Institute for Occupational Safety and Health (NIOSH) general guideline areas for reducing robot safety problems.

- Incorporating appropriate barriers with gates containing appropriate electrical interlocks so that the operation of a robot stops instantly in the event the gate is opened.
- Incorporating appropriate remote "diagnostic" instrumentation as much as possible so that most of the system trouble-shooting can be carried out from locations well outside the robot's operating zone.

The area, Worker training, is concerned with providing training to persons operating, maintaining, or programming robots. This training, in addition to emphasis on safe work-related practices, must stress factors such as [10,25]:

- Operators must not be in reach of a functioning robot.
- All involved persons must be familiar with all working aspects of the robot including safety barriers, emergency stops, known hazards, full range of motion and the programming of the robot, before operating or performing maintenance at robotic workstation.
- All operating, maintenance, and programming personnel must be aware of all possible pinch points, such as walls, poles, and other equipment, in the robot's operational zone.

Finally, the area, Worker supervision, contains guidelines such as [10,25]:

- Accept the fact that over a time period, even highly experienced persons carrying out automated tasks may become inattentive, over confident, or complacent to the hazards associated with automated systems.
- Do not permit individuals to enter operational areas of a robot without first putting the robot on "hold" mode, in a "power down" condition, or at a reduced operating speed mode.

8.8 Software Safety-Related Facts, Figures, and Examples

Some of the facts, figures, and examples directly or indirectly concerned with software safety include:

- Over 70% of the companies involved in the software development business develop their software by using ad hoc and unpredictable approaches/methods [26].
- The software industry in the United States is worth at least $300 billion per annum [27].

- An instrument failure due to a safety-related software issue caused the SAAB JAS39 Gripen fighter plane to crash [28].
- A radioactive heavy water spill at a Nuclear Power Generating Station in Canada was the result of a software error [28].
- Software errors in a computer-controlled therapeutic radiation machine known as Therac 25 resulted in deaths of two patients and severe injuries to another patient [29–32].
- A software error in a French meteorological satellite led to the destruction of 72 weather balloons [33].
- During the Gulf War, a software error shut down a Patriot missile system. Consequently, an enemy SCUD missile killed and wounded 124 persons [10,28].

8.9 Ways That Software Can Contribute to Hazards

There are many ways in which software can cause/contribute to a hazard. The main ones are [33,34]:

- Poor response to a contingency
- Poor timing of response for an adverse situation
- Failure to recognize a hazardous condition requiring a corrective action
- Carried out an unnecessary function
- Failure to carry out a necessary function
- Provided wrong solution to a problem
- Carried out a required function out-of-sequence

8.10 Basic Software System Safety-Related Tasks

Although, there are a large number of software system safety-related tasks, some of the basic ones include [35]:

- Clearly show the software system safety-related constraint consistency in regard to the software requirements specification.
- Trace all types of safety requirements and constraints right up to the code.

- Establish appropriate safety-related software test plans, test case requirements, test procedures, and test descriptions.

- Review with care the test results concerning safety issues and trace the identified safety-related software problems right back to the system level.

- Perform any special safety-related analyses, e.g., computer–human interface analysis or software fault tree analysis.

- Trace all identified system hazards to the hardware–software interface.

- Highlight with care all the safety-critical elements and variables for use by code developers.

- Identify the components of the software that control safety-critical operations and then direct all necessary safety analysis and tests on those particular functions and on the safety-critical path that leads to their execution.

- Develop on the basis of identified software system safety constraints the system-specific software design criteria and requirements, computer-human interface related requirements, and testing requirements.

- Develop a tracking system within the software framework along with system configuration control structure to assure the traceability of safety requirements and their flow through documentation.

8.11 Software Safety Assurance Program and Useful Software Safety Design-Related Guidelines

A software safety assurance program within the organization basically involves three maturity levels: I, II, and III [28]. Maturity level I is concerned with the development of company culture being clearly aware of software safety-related issues. More specifically, in this case, all software development personnel work according to standard development rules and apply them consistently.

Maturity level II is concerned with the implementation of a development process involving safety assurance reviews and hazard analysis for identifying and eliminating safety-critical conditions before being designed into the system. Finally, Maturity level III is concerned with the utilization of a design process that documents the results and implements continuous improvement methods to eradicate safety-critical errors out of the system software.

Nonetheless, some of the items to be considered during the implementation of a software safety assurance program include [10,28]:

- Software system safety is quantifiable to the specified risk level using the normal measuring methods.
- Past software safety data are fully considered and utilized in all potential software development projects.
- Software system safety-associated requirements are developed and specified as a component of the organization's design policy.
- All software system safety-related requirements are consistent with contract requirements.
- All human–computer interface-associated requirements are consistent with contract requirements.
- Software system safety is addressed in terms of a team effort that clearly involves groups such as engineering, quality assurance, and management.
- Changes in design, configuration, or requirements are carried out such that they still maintain an acceptable level of risk.
- All software system hazards are identified, evaluated, tracked, and eradicated as per requirements.

Finally, it is added that a software safety assurance program also must consider factors such as assuring that safety is designed into the system timely and cost-effectively, recording of safety data, and minimizing risk when using and accepting items, such as new materials, designs, and production and test methods [10,28].

Over the years, software development professionals have developed various useful software design-related guidelines. These guidelines are considered quite useful to improve software safety. Some of the guidelines are listed below [36]:

- Prohibit safety-critical software patches during the development process and remove unnecessary or obsolete code.
- Incorporate provisions to detect and log system errors.
- Ensure that conditional statements are under full software control and meet all possible conditions.
- Develop software design in such a way that it clearly prevents inadvertent/unauthorized access and/or modification to the code.
- Aim to separate and isolate all types of nonsafety-critical software modules from safety-critical modules.
- Avoid using all "0" and "1" digits for critical variables.

- Incorporate appropriate mechanism that causes the system to detect inadvertent jumps into or within safety critical computer software parts, and return itself to a state considered safe.
- Incorporate appropriate mechanism to ensure safety-critical computer software parts and interfaces to be under positive control at all times.
- Develop appropriate software modules to monitor critical software in regard to hazardous states, faults, errors, or timing-related problems.
- Incorporate the requirement for a password along with confirmation before the execution of a safety-critical software module.
- Incorporate an operator to validate or authorize the execution of safety-critical commands.

8.12 Software Hazard Analysis Methods

There are many methods that can be used to perform various types of software hazard analysis. Most of these methods are listed below [10,37–40]:

- Software sneak circuit analysis
- Code walk-through
- Event tree analysis
- Software fault tree analysis
- Proof of correctness
- Petri net analysis
- Hazard and operability studies
- Failure modes and effect analysis
- Cause-consequence diagrams
- Desk checking
- Monte Carlo simulation
- Cross reference-listing analysis
- Nuclear safety cross-check analysis
- Software/hardware integrated critical path
- Design walk-through
- Structural analysis

The first five of the above methods are described below [10,37–40].

8.12.1 Software Sneak Circuit Analysis

This method is used to identify software logic that causes undesired outputs. More specifically, program source code is converted to topological network trees and the code is modeled by using six basic patterns: iteration loop, parallel line, entry dome, trap, return dome, and single line. All software modes are modeled using the basic patterns linked in a network tree flowing right from top to bottom.

The analyst asks questions on the use and interrelations of the instructions considered elements of the structure. The effective answers to questions asked are useful in providing clues that highlight sneak conditions (an unwanted event not caused by component failure) that may lead to undesirable outputs. The analyst searches for the following four basic software sneaks:

1. Incorrect timing
2. A program message that poorly describes the actual condition
3. The undesirable inhibit of an output
4. Presence of an undesired output

All the clue-generating questions are taken from the topograph representing the code segment and at the discovery of sneaks, the analysts carry out investigative analyses to verify that the code does indeed generate the sneaks. Subsequently, all the possible impacts of the sneaks are assessed and necessary corrective actions recommended.

8.12.2 Code Walk-Through

This is a quite useful method to improve safety and quality of software products. The method is basically team effort among professionals, such as software engineers, program managers, software programmers, and the system safety professionals. Code walk-throughs are in-depth reviews of the software in process through discussion and inspection of the software functionality. All logic branches and each statement's function are discussed with care at a significant length. More simply, this process provides a good check and balances system of the software developed.

The system reviews the functionality of software and compares it with the specified system requirements. This provides a verification that all specified software safety requirements are implemented properly, in addition to the determination of functionality accuracy. Additional information on this method is available in Ref. [40].

8.12.3 Event Tree Analysis (ETA)

This method models the sequence of events resulting from a single initiating event. In regard to the application of this method to software, the initiating event is taken from a code segment considered safety critical (i.e., suspected of error or code inefficiencies). Usually, ETA assumes that each sequence event is either a failure or a success.

Some of the important factors associated with this method are listed below [22].

- The method always leaves some room to miss important initiating events.
- It is quite difficult to incorporate delayed recovery or success events.
- Normally, the method is used to perform analysis of more complex systems than the ones handled by the failure modes and effect analysis (FMEA) approach.
- It is a very good tool to identify undesirable events that require further investigation using the fault tree analysis (FTA) method.

Additional information on ETA is available in Refs. [22,41].

8.12.4 Software Fault Tree Analysis (SFTA)

This method is an offshoot of the fault tree analysis (FTA) method developed in the early 1960s at the Bell Telephone Laboratories to analyze the Minuteman Launch Control System from the safety aspect [42]. SFTA is used to analyze software design safety and its main objective is to demonstrate that the logic contained in the software design will not cause system safety failures, in addition to determining environmental conditions that may lead to the software causing a safety failure [43].

SFTA proceeds in a similar manner to hardware FTA described in Chapter 4. SFTA also highlights software–hardware interfaces. Although fault trees for both hardware and software are developed quite separately, they are linked together at their interfaces to allow total system analysis. This is very important because it is impossible to develop software safety procedures in isolation, but must be considered as a part of the total system safety.

Finally, it is added that although SFTA is an excellent hazard analysis method, it is a quite expensive tool to use. Additional information on FTA is available in Chapter 4 and in Refs. [22,42].

8.12.5 Proof of Correctness

This is a quite useful method to perform software hazard analysis. It decomposes a program under consideration into a number of logical segments and for each segment input/output assertions are defined. Subsequently, the involved software professional performs verification from the perspective that each and every input assertion and its associated output assertion are true and that, if all of the input assertions are true, then all of the output assertions also are true.

Finally, it is added that the proof of correctness approach makes use of mathematical theorem proving concepts to verify that a given program is clearly consistent with its associated specifications. Additional information on this method is available in Refs. [37–39].

Problems

1. Write an essay on robot and software safety.
2. What are the main causes of robot hazards?
3. Discuss safety considerations in robot life cycle.
4. Discuss human factors issues in robotic safety.
5. Describe at least four robot safeguard methods.
6. Discuss guidelines for reducing robot safety problems.
7. What are the main software hazard causing ways?
8. List at least 10 basic software system safety-related tasks.
9. Describe the following software hazard analysis methods:
    a. Software sneak circuit analysis
    b. Code walk-through
    c. Software fault tree analysis
10. Describe the software safety assurance program.

References

1. ANSI/RIA R15.06-1986, American National Standard for Industrial Robots: Safety Requirements, American National Standards Institute (ANSI), New York, 1986.
2. An Interpretation of the Technical Guidance on Safety Standards in the Use, etc., of Industrial Robots, Japanese Industrial Safety and Health Association, Tokyo, 1985.
3. Dhillon, B. S., Fashandi, A. R. M., Liu, K .L., Robot Systems Reliability and Safety: A Review, *Journal of Quality in Maintenance Engineering*, Vol. 8, No. 3, 2002, pp. 170–212.
4. Bonney, M. C., Yong, Y. F., Eds., *Robot Safety*, Springer-Verlag, New York, 1985.
5. Dhillon, B. S., *Robot Reliability and Safety*, Springer-Verlag, New York, 1991.
6. Cha, S. S., *Management Aspects of Software Safety*, Proceedings of the 8th Annual Conference on Computer Assurance, 1993, pp. 35–40.
7. Nagamachi, M., Ten Fatal Accidents Due to Robots in Japan, in *Ergonomics of Hybrid Automated Systems*, eds. W. Karwowski, et al., Elsevier, Amsterdam, 1988, pp. 391–396.

8. Sanderson, L. M., Collins, J. N., McGlothlin, J. D., Robot-Related Fatality Involving a U.S. Manufacturing Plant Employee: Case Report and Recommendations, *Journal of Occupational Accidents*, Vol. 8, 1986, pp. 13–23.

9. Hetzler, W. E., Hirsh, G. L., *Machine Operator Crushed by Robotic Platform*, Nebraska Fatality Assessment and Control Evaluation (FACE) Investigation Report No. 99NE017, The Nebraska Department of Labor, Omaha, NE, October 25, 1999.

10. Dhillon, B. S., *Engineering Safety: Fundamentals, Techniques, and Applications*, World Scientific Publishing, River Edge, NJ, 2003.

11. *Study on Accidents Involving Robots*, Report No. PB 83239822, Prepared by the Japanese Ministry of Labor, Tokyo, 1982. Available from the National Technical Information Service (NTIS), Springfield, VA.

12. Karwowski, W., Parsei, H. R., Amarnath, B., Rahimi, M., A Study of Worker Intrusion in Robots Work Envelope, in *Safety, Reliability, and Human Factors in Robotic Systems*, ed. J. H. Graham, Van Nostrand Reinhold, New York, 1991, pp. 148–162.

13. Jiang, B. C., Gainer, C. A., A Cause and Effect Analysis of Robot Accidents, *Journal of Occupational Accidents*, Vol. 9, 1987, pp. 27–45.

14. Clark, D. R., Lehto, M. R., Reliability, Maintenance, and Safety of Robots, in *Handbook of Industrial Robotics*, ed. S. Y. Nof, John Wiley & Sons, New York, 1999, pp. 717–753.

15. Backtrom, T., Dooes, M., A Comparative Study of Occupational Accidents in Industries with Advanced Manufacturing Technology, *International Journal of Human Factors in Manufacturing*, Vol. 5, 1995, pp. 267–282.

16. Industrial Robots and Robot System Safety, Chap. 4, in *OSHA Technical Manual*, Occupational Safety and Health Administration (OSHA), Department of Labor, Washington, D.C., 2001.

17. Ziskovsky, J. B., Working Safely with Industrial Robots, *Plant Engineering*, May 1984, pp. 81–85.

18. Russell, J. W., Robot Safety Considerations: A Checklist, *Professional Safety*, December 1983, pp. 36–37.

19. Nicolaisen, P., *Ways of Improving Industrial Safety for the Programming of Industrial Robots*, Proceedings of the 3rd International Conference on Human Factors in Manufacturing, November 1986, pp. 263–276.

20. Blache, K. M., Industrial Practices for Robotic Safety, in *Safety, Reliability, and Human Factors*, ed. J. H. Graham, Van Nostrand Reinhold, New York, 1991, pp. 34–65.

21. Zimmers, E. W., *Human Factors Aspects of Robotic Safety*, Proceedings of the Robotic Industries Association (RIA) Robot Safety Seminar, Chicago, April 24, 1986, pp. 1–8.

22. Dhillon, B. S., *Design Reliability: Fundamentals and Applications*, CRC Press, Boca Raton, FL, 1999.

23. Fox, D., Robotic Safety, *Robotics World*, January/February 1999, pp. 26–29.

24. Ulrich, K. T., Tuttle, T. T., Donoghue, J. P., Townsend, W. T., *Intrinsically Safer Robots*, NASA Report No. NAS 10-12178, Barrett Technology, Inc., Cambridge, MA, May 1995.

25. DHHS (NIOSH) Publication No. 85-103, *Preventing the Injury of Workers by Robots*, National Institute for Occupational Safety and Health (NIOSH), Morgantown, WV, 1984.

26. Thayer, R. H., Software Engineering Project Management, in *Software Engineering*, eds. M. Dorfman, R. H. Thayer, IEEE Computer Society Press, Los Alamitos, CA, 1997, pp. 358–371.
27. Hopcroft, J. E., Kraft, D. B., Sizing the U.S. Industry, *IEEE Spectrum*, December 1987, pp. 58–62.
28. Mendis, K. S., Software Safety and Its Relation to Software Quality Assurance, in *Handbook of Software Quality Assurance*, eds. G. G. Schulmeyer, J. I., McManus, Prentice Hall, Upper Saddle River, NJ, 1999, pp. 669–679.
29. Schneider, P., Hines, M. L. A., *Classification of Medical Software*, Proceedings of the IEEE Symposium on Applied Computing, 1990, pp. 20–27.
30. Gowen, L. D., Yap, M. Y., *Traditional Software Development's Effects on Safety*, Proceedings of the 6th Annual IEEE Symposium on Computer-Based Medical Systems, 1993, pp. 58–63.
31. Joyce, E., Software Bugs: A Matter of Life and Liability, *Datamation*, Vol. 33, No. 10, 1987, pp. 88–92.
32. Dhillon, B. S., *Medical Device Reliability and Associated Areas*, CRC Press, Boca Raton, FL, 2000.
33. Leveson, N. G., Software Safety: Why, What, and How, *Computing Surveys*, Vol. 18, No. 2, 1986, pp. 125–163.
34. Friedman, M. A., Voas, J. M., *Software Assessment*, John Wiley & Sons, New York, 1995.
35. Leveson, N. G., *Software*, Addison-Wesley Publishing Company, Reading, MA, 1995.
36. Keene, S. J., *Assuring Software Safety*, Proceedings of the Annual Reliability and Maintainability Symposium, 1992, pp. 274–279.
37. Ippolito, L. M., Wallace, D. R., *A Study on Hazard Analysis in High Integrity Software Standards and Guidelines*, Report No. NISTIR 5589, National Institute of Standards and Technology, U.S. Department of Commerce, Washington, D.C., January 1995.
38. Hammer, W., Price, D., *Occupational Safety Management and Engineering*, Prentice Hall, Upper Saddle River, NJ, 2001.
39. Hansen, M. D., *Survey of Available Software-Safety Analysis Techniques*, Proceedings of the Annual Reliability and Maintainability Symposium, 1989, pp. 46–49.
40. Sheriff, Y. S., Software Safety Analysis: The Characteristics of Efficient Technical Walk-Throughs, *Microelectronics and Reliability*, Vol. 32, No. 3, 1992, pp. 407–414.
41. Cox, S. J., Tait, N. R. S., *Reliability, Safety, and Risk Management*, Butterworth-Heinemann Ltd., London, 1991.
42. Dhillon, B. S., Sing, C., *Engineering Reliability: New Techniques and Applications*, John Wiley & Sons, New York, 1981.
43. Leveson, N. G., Harvey, P. R., Analyzing Software Safety, *IEEE Transactions on Software Engineering*, Vol. 9, No. 5, 1983, pp. 569–579.

9

Human Error in Transportation Systems

9.1 Introduction

Each year, billions of dollars are spent around the world to develop, manufacture, operate, and maintain transportation systems, such as trains, motor vehicles, aircraft, and ships. During the day-to-day use of such systems, thousands of lives are lost worldwide each year due to accidents resulting from various types of problems, including human errors.

The railway system is still an important mode of transportation around the globe. For example, in the United States, the railway system is made up of about 3000 stations and track terminals serving roughly 15 large freight railroads and over 600 small, regional roads. Nonetheless, over the years around the globe, a large number of fatal accidents have occurred due to various types of human errors in the railway system [1–3].

In road transportation systems, safety has become a pressing issue because around the world approximately 0.8 million road accident fatalities and 20 to 30 million injuries occur each year [4,5]. Human error is believed to be an important factor in the occurrence of such events.

In the area of aviation, although the overall accident rate for air travel has declined quite considerably over the years, the reduction in human error-related accidents in aviation has failed to keep pace with the reduction of accidents resulting from mechanical and environmental factors [6–8].

In the area of shipping, although many of the systems used in a modern ship may be fully automated, they still require a degree of human intervention (e.g., respond to alarms or set initial tolerances). Needless to say, as per past experiences, about 80% of accidents in the shipping industry are rooted in human error [9].

This chapter presents various important aspects of human error in transportation systems.

9.2 Railway System Human Error-Related Facts and Figures

Some of the railway system's directly or indirectly human error-related facts and figures are as follows:

- In Norway, during the period from 1970 to 1998, about 62% of the 13 railway accidents that caused fatalities or injuries were due to human errors [10].
- In the United Kingdom, during the period from 1900 to 1997, about 70% of the 141 accidents on four British Railway main lines were due to human errors [10,11].
- In the United States, during the period from 2000 to 2004, train accidents caused 4,623 fatalities and 51,597 injuries [12].
- In 2004, in the United States about 53% of the railway switching yard accidents (i.e., excluding highway-rail crossing train accidents) were due to human factors-related causes [1].
- In 1999, 31 people died and 227 persons were hospitalized in train accidents in the United Kingdom caused by a human error [13].
- In India, each year over 400 railway accidents occur and about 66% of these accidents are directly or indirectly due to human error-related causes [14].
- In 2005, a subway train crash at Thailand Cultural Center Station, Bangkok, due to a human error injured about 200 persons [15].

9.3 Railway Personnel Tasks Prone to Serious Human Error

Railway personnel perform a wide variety of tasks during their day-to-day work environment. Past experiences indicate that some of the tasks performed by the railway personnel are more prone to human error than the others. Nonetheless, some of the tasks performed by the railway personnel in their day-to-day work environment that are prone to the occurrence of serious human errors include [13]:

- Railway personnel maintaining the tracks
- Engine-driver driving the train
- Railway personnel loading the wagons
- Switching person controlling the points
- Railway personnel maintaining the concerned vehicles
- Railway personnel maintaining the track systems/devices

- Station foreman or supervisor receiving the train into the railway station
- Station foreman or supervisor dispatching the train from the railway station

9.4 Typical Human Error Occurrence Areas in Railway Operation

In railway operation, there are many areas for the occurrence of human errors. The three typical ones are shown in Figure 9.1 [10]. These are train speed, signal passing, and signaling or dispatching.

In the area of train speed, in the past many railway accidents have occurred because of the failure of train drivers to reduce speed of the train as specified for the route in question. The likelihood of over speeding and its related consequences depend on many factors including the type of speed restrictions and the circumstances surrounding it.

Basically, there are three types of speed restrictions that require driver response from his or her perspective:

1. **Conditional speed restrictions.** These types of speed restrictions are imposed because of train route setting at a station or junction and the signaling aspect displayed in that regard.
2. **Emergency or temporary speed restrictions.** These types of speed restrictions are imposed because of temporary track deficiencies, such as stability problems and frost heave, or maintenance work.

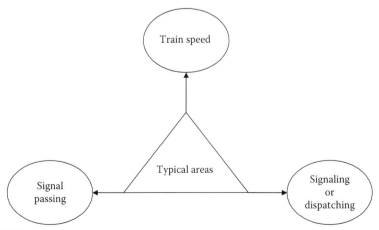

FIGURE 9.1
Three typical areas for the occurrence of human errors in railway operation.

3. **Permanent speed restrictions.** These types of speed restrictions are imposed because of track curves or some existing infrastructure-related conditions on a specific section of a track in question.

In the area of signal passing, in the past, many railway accidents have occurred. Trains passing a signal displaying a stop is a dangerous occurrence because it can result in an immediate conflict with another train or trains. This event or situation is frequently referred to as Signal Passed at Danger (SPAD). Each year, many SPAD-related incidents occur around the globe. For example, the figure for the period 1996 to 1997, for the British Railway System alone, was 653 [10]. Some of the main causes for the occurrence of a SPAD event include [10,16]:

- Oversight or disregard of a signal
- Misjudging of which signal applies to the train in question
- Failure to see signal because of poor visibility
- Over speeding with respect to braking performance and warning signal distance
- Misunderstanding of signaling aspect
- Misjudging the effectiveness of brakes under particular circumstances, such as bad weather
- Driver falls asleep or is unconscious

In the area of signaling or dispatching, in the past, many railway accidents have occurred because of errors made by dispatchers or signalmen. Nowadays, with the use of modern technical devices, human errors in this area have been reduced quite significantly [10].

9.5 A Useful Checklist of Statements to Reduce Human Error in Railways

This section presents a checklist of statements to ensure good human factor practices in railway-related projects. In turn, this exercise will be useful to reduce human error occurrences in the railway system at large. The list is presented below [17].

- Human reliability analysis methods and techniques are used properly and effectively.
- Individuals performing human factors-related tasks are competent to do so.
- The human error identification process is fully integrated into the general hazard identification process, within the framework of the project.

- All human errors are identified, modeled, and controlled correctly and effectively.
- Human factors are given the same degree of importance as any other area of safety engineering.
- All human factors-related aspects are being considered from the outset of a given project.
- Individuals involved in performing human factors-related tasks are given sufficient resources and authority.
- A broad range of human factors-related information is being communicated effectively.
- The project aims to design systems that help all types of potential users avoid or recover from hazards.
- The representation of human error is fully integrated effectively with other aspects of safety analysis.
- All necessary aspects of human factors are integrated effectively into the safety argument.
- The tasks being performed are clearly understood in order to highlight sources of human error.
- The identification, evaluation, and reduction of risk from the occurrence of human errors is being considered as an important element of any safety process.
- All the human factors planning aspects are integrated into the general project planning.
- The required and existing competency of all types of end users is evaluated in an effective manner.
- All types of human factors-related requirements are fully integrated in an effective manner into the system requirements.
- When considering risk reduction methods, all types of potential users are involved.

9.6 Road Transportation Systems Human Error-Related Facts and Figures

Some of the road transportation systems', directly or indirectly, human error-related facts and figures are as follows:

- The annual cost of worldwide road crashes is estimated to be about $500 billion and by the year 2020, it is predicted that road traffic injuries will become the third largest cause of disabilities in the world [18–20].

- During the period 1966 to 1998, over 70% of bus-related accidents were the result of driver error in five developing countries: India, Zimbabwe, Thailand, Tanzania, and Nepal [4,16].
- Human error is cited more frequently than mechanical-related problems in about 5000 truck-related fatalities that occur annually in the United States [21].
- As per Ref. [22], around 65% of motor vehicle-related accidents are attributable to human error.
- A study of heavy duty truck accidents reported that around 80% of the accidents were the result of human error [23].
- A study of truck–car crashes reported that most of these crashes were the result of human error either committed by the car driver or truck driver [24].
- As per Ref. [25], about 57% of bus accidents in South Africa are the result of human error.

9.7 Operational Influences on Commercial Driver Performance and Classifications of Driver Errors

Operational influences on the performance of commercial drivers play an important role in regard to the occurrence of human error. Although all drivers carry out their tasks in a motor vehicle operating within the physical environment of a road, a major difference lies in the commercial vehicle transportation operational environment. More specifically, commercial drivers perform their tasks against the backdrop of a complex operational environment that includes items such as [24]:

Work-related requirements. A typical example of such requirements is customer delivery schedules.

Practices stated by the company management. These practices include items such as scheduling, training, and incentives for safe work performance.

Labor policies and traditions.

Government or other body regulations and violation penalties.

Finally, it is suggested that, to the greatest extent possible, the operational environment must optimize safety in regard to the occurrence of human errors while sustaining an acceptable level of productivity.

Drivers make various types of errors. They may be grouped, in regard to the decreasing frequency of occurrence, under the following four classifications [26,27]:

- Recognition errors
- Decision errors
- Performance errors
- Miscellaneous errors

Additional information on these classifications is available in Refs. [26,27].

9.8 Common Driver Errors and Ranking of Driver Errors

Drivers make various types of errors during the driving process. The common driver errors include [28,29]:

- Changing lanes abruptly
- Following too closely
- Overtaking at junction or cross road
- Following closely a motor vehicle that is overtaking
- Passing or overtaking in the face of incoming traffic
- Following closely before overtaking
- Driving too fast for prevailing circumstances
- Straddling lanes

Over the years various studies have been conducted to rank the occurrence of driver errors. The results of one of these studies that ranked driver errors/causes from highest frequency of occurrence to the lowest frequency of occurrence are presented in Table 9.1 [28].

9.9 Aviation Systems Human Error-Related Facts and Figures

Some of the aviation systems human error-related facts and figures include:

- A Boeing study reported that the failure of the cockpit crew has been a contributing factor in the occurrence of over 73% of aircraft accidents worldwide [30,31].

TABLE 9.1

Ranking of Driver Errors From Highest Frequency of
Occurrence to the Lowest Frequency of Occurrence

Error Rank (Highest Occurrence to Lowest Occurrence)	Error Description
1	• Lack of care
2	• Too fast
3	• Looked, but failed to see
4	• Distraction
5	• Inexperience
6	• Failure to look
7	• Incorrect path
8	• Poor attention
9	• Improper overtaking
10	• Wrong interpretation
11	• Lack of judgment
12	• Misjudged distance and speed
13	• Following too closely
14	• Difficult maneuver
15	• Reckless or irresponsible
16	• Incorrect decision/action
17	• Lack of education or road craft
18	• Faulty signaling
19	• Poor skill

- As per Ref. [32], during the period 1983 to 1996, there were 371 major airline crashes, 1,735 commuter/air taxi crashes, and 29,798 general aviation crashes. A study of these crashes reported that pilot error was a probable cause for about 38% of major airline crashes, 74% of commuter/air taxi crashes, and 85% of general aviation crashes [32].

- A study reported that since the introduction of highly reliable turbojet aircraft in the late 1950s, over 70% of airline accidents involved some degree of human error [33].

- A study reported that about 45% of major airline crashes occurring at airports are due to pilot error as opposed to about 28% of those occurring elsewhere [32].

- During the period 1990 to 1996, a study reported that pilot error was responsible for about 34% of major airline crashes [32].

- A study reported that crashes due to pilot error in major airlines in the United States decreased from 43% for the period 1983 to 1989 to 34% for the period 1990 to 1996 [16,32].

9.10 Contributory Factors to Flight Crew Decision Errors

There are many factors that can contribute to flight crew decision errors with respect to incidents. In particular, at minimum the factors that must be assessed with care in regard to their significance in contributing to flight crew decision-related errors are shown in Figure 9.2 [34]. These are crew factors, the procedure from which the error resulted, equipment factors, flight phase where error occurred, environmental factors, and other stimuli (i.e., beyond indications). The crew factors include items such as intention, personal and corporate stressors, technical knowledge/skills/experience, crew communication/coordination, crew understanding of situation at the time of procedure execution, situation awareness factors (e.g., attention, vigilance, etc.), and factors that affect individual performance. (e.g., workload, fatigue, etc.)

The procedure from which the error resulted includes items such as crew interpretation of the relevant procedure, current guidelines and policies aimed at prevention of incident, onboard source of the procedure, procedural factors (e.g., impracticality, complexity, negative transfer, etc.), and procedure status. The equipment factors include items such as airplane configuration, airplane flight deck indications, and the role of automation.

The remaining three contributory factors, i.e., flight phase where error occurred, environmental factors, and other stimuli (i.e., beyond indications), are considered self-explanatory, but the additional information on these factors is available in Ref. [34].

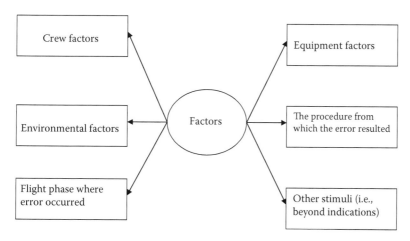

FIGURE 9.2
Contributory factors to flight crew decision errors.

9.11 Types of Pilot-Controller Communication-Related Errors and Useful Recommendations for Reducing Them

The communication between pilots and air traffic controllers is very important and is subject to various types of errors. A study of 386 reports submitted to the Aviation Safety Reporting System (ASRS) during the period from July 1991 to May 1996 indicates that pilot–controller communication-related errors may be grouped under the following four types [35]:

1. **Readback/hearback errors.** These types of errors, as per the study, accounted for 47% of the pilot–controller communication-related errors and the most common contributing factor for their occurrence was similar call signs [35].

2. **No pilot readback.** This accounted for, as per the study, 25% of the pilot–controller communication-related errors and the most common contributing factor for its occurrence was the pilot expectation [35].

3. **Hearback errors type II.** These are types of errors in which a pilot correctly repeats the issued clearance, but the controller overlooks the issued clearance that, in fact, was not the clearance he/she intended to issue. It is to be noted that this type of error also incorporates events where the pilot made a problematic action statement or intent that the controller should have noticed. As per the study, hearback errors type II accounted for 18% of the pilot–controller communication-related errors [35].

4. **Miscellaneous errors.** These are types of errors that cannot be grouped under the above three types. As per the study, the miscellaneous errors accounted for 10% of the pilot–controller communication-related errors [35].

Some of the useful recommendations, for reducing communication-related errors between pilots and controllers are [35]:

- Encourage air traffic controllers to treat all types of readbacks as they would treat any other type of incoming information.
- Encourage all controllers to speak distinctly and slowly.
- Encourage controllers to avoid issuing "strings" of instructions to different aircraft.
- Encourage all aircraft pilots to respond to all controller instructions with complete readback of all important components.
- When having similar call signs on the frequency, encourage all involved aircraft pilots to say their call sign before and after each readback.

- When having similar call signs on the frequency, encourage all involved air traffic controllers to continue to announce this fact.
- Encourage air traffic controllers to keep all types of instructions short with a maximum of four instructions per transmission.

9.12 Organizational-Related Factors in Commercial Aviation Accidents in Regard to Pilot Error

Over the years various studies have been performed to identify organizational-related factors in commercial aviation accidents in regard to pilot error. One of these studies analyzed the National Transportation Safety Board's (NTSB) commercial aviation accident data for the period 1990 to 2000. The study reported that during this period 60 of the 1322 accidents were attributable, directly or indirectly, to pilot error due to 70 organizational-related causes [36]. These causes or factors are grouped under the following 10 categories along with their corresponding brief descriptions in parentheses [36]:

Category I: Faulty documentation (i.e., wrong signoffs, checklists, and record keeping that effects flight operations).

Category II: Poor procedures or directives (i.e., ill-defined or conflicting policies, formal oversight of operation).

Category III: Inadequate surveillance of operations (i.e., organizational climate issues, chain-of-command, and quality assurance and trend information).

Category IV: Company/management induced pressures (i.e., threats to pilot job status and/or pay).

Category V: Poor facilities (i.e., failure to provide satisfactory environmental controls, lighting, clearance, etc., for flight operations).

Category VI: Insufficient or untimely information sharing (i.e., weather reports, logbooks, and updates on the part of the organization).

Category VII: Poor standards/requirements (i.e., adherence to policy and clearly defined organizational objectives).

Category VIII: Inadequate substantiation process (i.e., well-defined and verified process, accountability, standards of operation, regulation, and reporting/recording process).

Category IX: Inadequate initial, upgrade or emergency training/transition (i.e., opportunities for pilot training not implemented or made available to appropriate pilots).

Category X: Poor supervision of operations at management level (i.e., failure to provide necessary guidance, leadership to flight operations, and oversight).

The study also reported the percentage contributions of the organizational causes or factors belonging in each of the above 10 categories to the 60 accidents attributable, directly or indirectly, to pilot error. Thus, these 10 categories of the organizational causes or factors along with their corresponding percentage contribution to the 60 accidents are as follows [36]:

Category I: (faulty documentation): 4%

Category II: (poor procedures or directives): 21%

Category III: (inadequate surveillance of operations): 13%

Category IV: (company/management induced pressures): 6%

Category V: (poor facilities): 1.5%

Category VI: (insufficient or untimely information sharing): 12%

Category VII: (poor standards/requirements): 12%

Category VIII: (inadequate substantiation process): 3%

Category IX: (inadequate initial, upgrade, or emergency training/ transition): 18%

Category X: (poor supervision of operations at management level): 10%

9.13 Shipping Systems Human Error-Related Facts and Figures

Some of the shipping systems' directly or indirectly human error-related facts and figures are as follows:

- A study of 6091 accident claims over $100,000 concerning all classes of commercial ships over a period of 15 years, performed by the United Kingdom Protection and Indemnity (UKP&I) club, reported that about 62% of the claims were attributable to human error [37–39].

- A Dutch study of 100 marine casualties reported that human error was a factor in 96 of the 100 accidents [40,41].

- As per Refs. [40,42], human error contributes to 84 to 88% of tanker-related accidents.

- In 2004, *Bow Mariner*, a chemical/product tanker, sunk due to an onboard explosion caused by a human error and 18 crew members died [43].

- As per the findings of the UKP&I club, the occurrence of human errors costs the maritime industry $541 million per year [37].

- As per Refs. [44,45], over 80% of marine-related accidents are influenced or caused by human and organization factors.
- A human error caused the collision of the *MV Santa Cruz II* and the USCGC *Cuyahoga* that resulted in the deaths of 11 Coast Guardsmen [40,46].
- As per Refs. [40,47], human error is a factor in 89 to 96% of ship collisions.
- The grounding of the ship named *Torrey Canyon* due to various types of human errors caused the spilling of 100,000 tons of oil [40,46].
- As per Refs. [40,48], human error is a factor in 79% of towing vessel groundings.

9.14 Marine Industry Human Factors Issues

Nowadays, there are many human factors-related issues facing the marine industrial sector that can directly or indirectly influence the occurrence of human error. Some of these issues include [40,49–53]:

- **Faulty policies, standards, or practices.** This issue covers a variety of problems including the lack of standard traffic rules from port to port, the lack of available precisely written and comprehensible operational procedures aboard ship, and management policies that encourage risk-taking.
- **Poor communications.** This issue is concerned with communications between masters and pilots, shipmates, ships, etc. As per Ref. [52], about 70% of marine collisions and allisions occurred when a federal or state pilot was directing one or both vessels. In this regard, better procedures and training can be quite useful in promoting better communications and coordination on and between vessels.
- **Poor general technical knowledge.** This issue is concerned with the poor comprehension of mariners regarding how the automation functions or under what conditions it was designed to function effectively.
- **Fatigue.** This is a "pressing issue" of mariners as pointed out by two different studies [49,50]. Furthermore, another study reported that fatigue was a contributing factor in 33% of the vessel injuries and 16% of the casualties [53].
- **Decisions based on inadequate information.** This issue is concerned with mariners making navigation associated decisions on the basis of inadequate information. As per past experiences, they frequently tend to rely on either their memory or a favored piece of

equipment and, in other cases, essential information could be lacking or wrong altogether. Situations such as these can result in navigation errors.

- **Hazardous natural environment.** This issue is concerned with fog, winds, and currents that can make quite treacherous working conditions, thus a greater degree of risk for the occurrence of casualties. This problem could be overcome by seriously considering these three factors (i.e., fog, winds, and currents) during a ship's and equipment design process as well as making appropriate adjustments to ship operations on the basis of hazardous environmental conditions.

- **Poor automation design.** This is a quite challenging issue because poor equipment design pervades almost all types of shipboard automation. According to Ref. [41], poor equipment design was a causal factor in around 33% of major marine casualties. In this aspect, a careful consideration by equipment designers to factors, such as how a given piece of equipment will effectively support the tasks performed by mariners and how it will effectively integrate into the complete equipment "suite" used by mariners, can be a very useful step.

- **Poor knowledge of the ships' systems.** This is an important issue because it is a frequent factor in marine casualties because of difficulties encountered by pilots and crews working on ships of different sizes, with a varying variety of equipment and carrying different types of cargoes. Actions, such as standardized equipment design, better training, and an effective approach to assigning crews to ships, can be very helpful in overcoming this difficulty.

- **Poor maintenance.** This is a challenging issue because poor maintenance of ships can result in situations, such as lack of functional backup systems, crew fatigue from the need to perform emergency repairs, and dangerous work environments. As per Ref. [47], poor maintenance is a leading cause of explosions and fires in ships.

9.15 Approaches to Reduce the Manning Impact on Shipping System Reliability

Three useful approaches to reduce the manning impact on shipping system reliability with respect to improving human reliability are as follows [54]:

Approach I: Eliminate or minimize impacts of human error. In this case, the impacts of human error are minimized or eliminated altogether through actions such as designing the system to be fully error tolerant, and designing the system that clearly enables human/

system to recognize that an error has occurred and to correct the error prior to the occurrence of any damage.

Approach II: Reduce the human error incidence occurrence. In this case, human error rates are reduced through actions such as job task simplification, application of human engineering design principles, and error occurrence likelihood analysis or modeling.

Approach III: Improve mean time between failures under the reduced manning environment. In this case, mean time between failures is improved through actions such as designing or choosing highly reliable system parts/components and designing the interfaces to optimize the use of these parts/components.

Additional information on this topic is available in Ref. [54].

Problems

1. Write an essay on human error in transportation systems.
2. List at least six railway system human error-related facts and figures.
3. What are the railway personnel day-to-day tasks prone to serious human error?
4. Discuss the typical human error occurrence areas in railway operation.
5. What are the common driver errors?
6. List at least five aviation systems human error-related facts and figures.
7. What are the main contributory factors to flight crew decisions errors?
8. List the useful recommendations to reduce communication-related errors between pilots and controllers.
9. List at least 10 shipping systems human error-related facts and figures.
10. What are the important marine industry human factors issues that can directly or indirectly influence the occurrence of human error?

References

1. Reinach, S., Viale, A., Application of a Human Error Framework to Conduct Train Accident/Incident Investigations, *Accident Analysis and Prevention*, Vol. 38, 2006, pp. 396–406.

2. Report No. DOT/FRA/RRS-22, Federal Railroad Administration (FRA) Guide for Preparing Accident/Incident Reports, FRA Office of Safety, Washington, D.C., 2003.
3. Wittingham, R. B., *The Blame Machine: Why Human Error Causes Accidents*, Elsevier Butterworth-Hinemann, Oxford, U.K., 2004.
4. Pearce, T., Maunder, D. A. C., *The Causes of Bus Accidents in Five Emerging Nations*, Report, Transport Research Laboratory, Wokingham, U.K., 2000.
5. Jacobs, G., Aeron-Thomas, A., Astrop, A., *Estimating Global Road Fatalities*, Report No. TRL 445, Transport Research Laboratory, Wokingham, U.K., 2000.
6. Wiegmann, D. A., Shappell, S. A., *A Human Error Approach to Aviation Accident Analysis*, Ashgate Publishing Limited, London, U.K., 2003.
7. Report No. PB94-917001, *A Review of Flight Crew Involved in Major Accidents of U.S. Air Carriers, 1978–1990*, National Transportation Safety Board, Washington, D.C., 1994.
8. Nagel, D., Human Error in Aviation Operations, in *Human Factors in Aviation*, eds. E. Wiener and D. Nagel, Academic Press, San Diego, CA, 1988, pp. 263–303.
9. Fotland, H., Human Error: A Fragile Chain of Contributing Elements, *The International Maritime Human Element Bulletin*, No. 3, April 2004, pp. 2–3.
10. Anderson, T., *Human Reliability and Railway Safety*, Proceedings of the 16th European Safety, Reliability, and Data Association (ESREDA) Seminar on Safety and Reliability in Transportation, 1999, pp. 1–12.
11. Hall, S., *Railway Accidents*, Ian Allan Publishing, Shepperton, U.K., 1997.
12. *Accident/Incident Tool*, Federal Railroad Administration (FRA) Office of Safety, Federal Railroad Administration, Washington, D.C., 2005.
13. Hudoklin, A., Rozman, V., Reliability of Railway Traffic Personnel, *Reliability Engineering and System Safety*, Vol. 52, 1996, pp. 165–169.
14. White Paper on Safety on Indian Railways, Railway Board, Ministry of Railways, Government of India, New Delhi, India, April 2003.
15. Human Error Derails New Metro, Editorial, *The Nation Newspaper*, Bangkok, Thailand, January 18, 2005.
16. Dhillon, B. S., *Human Reliability and Error in Transportation Systems*, Springer-Verlag, London, 2007.
17. *Human Error: Causes, Consequences, and Mitigations*, Application Note 3, Railway Safety, London, U.K., 2003.
18. Odero, W., *Road Traffic Injury Research in Africa: Context and Priorities*, paper presented at the Global Forum for Health Research Conference (Forum 8), November 2004. Available from the School of Public Health, Moi University, Eldoret, Kenya.
19. Krug, E., Ed., *Injury: A Leading Cause of the Global Burden of Disease*, World Health Organization (WHO), Geneva, Switzerland, 1999.
20. Murray, C. J. L., Lopez, A. D., *The Global Burden of Disease*, Harvard University Press, Boston, 1996.
21. Trucking Safety Snag: Handling Human Error, *The Detroit News*, Detroit, MI, July 17, 2000.
22. *Driving Related Facts and Figures*, U.K., July, 2006. Available online at www.drive-andsurvive.ca.uk/cont5.htm
23. Human Error to Blame for Most Truck Mishaps, UW Prof Says, *News Bureau*, University of Waterloo (UW), Ontario, Canada, April 18, 1995.

24. Zogby, J. J., Knipling, R. R., Werner, T. C., *Transportation Safety Issues*, Report by the Committee on Safety Data, Analysis, and Evaluation, Transportation Research Board, Washington, D.C., 2000.
25. Poor Bus Accident Record for Gauteng, *South African Press Association (SAPA)*, Cape Town, South Africa, July 4, 2003.
26. Rumar, K., The Basic Driver Error: Late Detection, *Ergonomics*, Vol. 33, 1990, pp. 1281–1290.
27. Treat, J. R., *A Study of Pre-crash Factors Involved in Traffic Accidents*, Report No. HSRI 10/11, 6/1, Highway Safety Research Institute (HSRI), University of Michigan, Ann Arbor, 1980.
28. Brown, I. D., Drivers' Margin of Safety Considered as a Focus for Research on Error, *Ergonomics*, Vol. 33, 1990, pp. 1307–1314.
29. Harvey, C. F., Jenkins, D., Sumner, R., *Driver Error*, Report No. TRRL SR 149, Transport and Research Laboratory (TRRL), Department of Transport, Crowthorne, U.K., 1975.
30. Report No. 1-96, *Statistical Summary of Commercial Jet Accidents: Worldwide Operations: 1959–1996*, Boeing Commercial Airplane Group, Seattle, Washington, 1996.
31. Mjos, K., Communication and Operational Failures in the Cockpit, *Human Factors and Aerospace Safety*, Vol. 1, No. 4, 2001, pp. 323–340.
32. Fewer Airline Crashes Linked to "Pilot Error," Inclement Weather Still Major Factor, *Science Daily*, January 9, 2001.
33. Helmreich, R. L., Managing Human Error in Aviation, *Scientific American,* May 1997, pp. 62–67.
34. Graeber, R. C., Moodi, M. M., *Understanding Flight Crew Adherence to Procedures: The Procedural Event Analysis Tool (PEAT)*, Proceedings of the Joint Meeting of the 51st FSF International Air Safety Seminar and the 28th IFA International Conference, 1998, pp. 415–424.
35. Cardosi, K., Falzarano, P., Han, S., *Pilot-Controller Communication Errors: An Analysis of Aviation Safety Reporting System (ASRS) Reports*, Report No. DOT/FAA/AR-98/17, Federal Aviation Administration (FAA), Washington, D.C., August 1998.
36. Von Thaden, T. L., Wiegmann, D. A., Shappell, S. A., *Organization Factors in Commercial Aviation Accidents 1990–2000*, paper presented at the 13th International Symposium on Aviation Psychology, Dayton, Ohio, 2005. Available from D. A. Wiegmann, University of Illinois at Urbana-Champaign, Champaign, IL.
37. Just Waiting to Happen—The Work of the UK P&I Club, *The International Maritime Human Element Bulletin*, No. 1, October 2003, pp. 3–4.
38. DVD Spotlights Human Error in Shipping Accidents, *Asia Maritime Digest,* January/February 2004, pp. 41–42.
39. Boniface, D. A., Bea, R. G., Assessing the Risks of and Countermeasures for Human and Organizational Error, *SNAME Transactions*, Vol. 104, 1996, pp. 157–177.
40. Rothblum, A. M., *Human Error and Marine Safety*, Proceedings of the Maritime Human Factors Conference, Linthicum, MD, USA, 2000, pp. 1–10.
41. Wagenaar, W. A., Groeneweg, J., Accidents at Sea: Multiple Causes and Impossible Consequences, *International Journal of Man–Machine Studies*, Vol. 27, 1987, pp. 587–598.
42. Working Paper on Tankers Involved in Shipping Accidents 1975–1992, Transportation Safety Board of Canada, Ottawa, Canada, 1993.

43. Human Error Led to the Sinking of the *Bow Mariner, The Scandinavian Shipping Gazette*, Gothenburg, Sweden, 2006.
44. Hee, D. D., Pickrell, B. D., Bea, R. G., Roberts, K. H., Williamson, R. B., Safety Management Assessment System (SMAS): A Process for Identifying and Evaluating Human and Organization Factors in Marine System Operations with Filed Test Results, *Reliability Engineering and System Safety*, Vol. 65, 1999, pp. 125–140.
45. Moore, W. H., Bea, R. G., *Management of Human Error in Operations of Marine Systems,* Report No. HOE-93-1, 1993. Available from the Department of Naval Architecture and Offshore Engineering, University of California, Berkeley.
46. Perrow, C., *Normal Accidents: Living with High Risk Technologies*, Basic Books, Inc., New York, 1984.
47. Bryant, D. T., *The Human Element in Shipping Casualties*, Report prepared for the Department of Transport, Marine Directorate, London, U.K., 1991.
48. Cormier, P. J., Towing Vessel Safety: Analysis of Congressional and Coast Guard Investigative Response to Operator Involvement in Casualties Where a Presumption of Negligence Exists, master's thesis, University of Rhode Island, Kingston, 1994.
49. *Crew Size and Maritime Safety*, Report by the National Research Council, National Academy Press, Washington, D.C., 1990.
50. *Human Error in Merchant Marine Safety*, Report by the Marine Transportation Research Board, National Academy of Science, Washington, D.C., 1976.
51. *Prevention through People: Quality Action Team Report*, U.S. Coast Guard, Washington, D.C., 1995.
52. *Major Marine Collisions and Effects of Preventive Recommendations*, Report by the National Transportation Safety Board (NTSB), Washington, D.C., 1981.
53. McCallum, M. C., Raby, M., Roghblum, A. M., *Procedures for Investigating and Reporting Human Factors and Fatigue Contributions to Marine Casualties*, U.S. Coast Guard Report No. CG-D-09-07, Department of Transportation, Washington, D.C., 1996.
54. Anderson, D. E., Malone, T. B., Baker, C. C., Recapitalizing the Navy through Optimized Manning and Improved Reliability, *Naval Engineers Journal*, November 1998, pp. 61–72.

10

Human Error in Healthcare Systems and in Mining Equipment

10.1 Introduction

Each year a vast sum of money is spent worldwide to produce various types of systems or devices for use in healthcare. For example, in 1997 the world market for medical devices was estimated to be around $120 billion [1]. The history of the use of medical devices may be traced back to the ancient Egyptians and Etruscans who used dental devices [2]. Today, a modern hospital uses more than 5000 types of medical devices, ranging from a complex pacemaker to a simple tongue depressor [3].

Over the years many deaths and injuries have occurred due to human error in healthcare systems or devices. For example, as per Ref. [4], deaths or serious injuries associated with medical devices reported through the U.S. Food and Drug Administration's (FDA) Center for Devices and Radiological Health (CDRH) accounted for around 60% attributed to user error. Nowadays, information on human error in healthcare systems is widely available in the form of book chapters, journal articles, conference proceedings articles, etc., as human error in healthcare systems has become a pressing issue [5,6].

In the past, the mining industry has played a pivotal role in the development of civilization, and today the world mining industry is producing over 6 billion tons of raw product annually, which is worth trillions of dollars. Each year billions of dollars are spent to produce various types of equipment for use by the mining industrial sector worldwide. For example, in the United States in 2004, mining equipment manufacturers shipped around $1.4 billion worth of goods [7].

Over the years, human errors in mining equipment have directly or indirectly resulted in deaths and serious injuries. For example, as per Ref. [8], 794 errors resulted in fatal accidents in mines.

Over the years, professionals working in the areas of healthcare systems and mining equipment have developed various types of methods and approaches to reduce the occurrence of human errors in these areas.

This chapter presents various important aspects of human error in health care systems and in mining equipment.

10.2 Healthcare Systems Human Error-Related Facts and Figures

Some of the facts and figures directly or indirectly concerned with human errors in healthcare systems are as follows:

- Human errors cause or contribute to up to 90% of accidents associated with medical devices [9–11].
- Each year the Food and Drug Administration (FDA) in the United States receives approximately 100,000 reports through the medical device reporting (MDR) route and 5,000 reports through the voluntary MedWatch program [12]. A significant proportion of these reports directly or indirectly are concerned with human factors-related problems.
- A hospital paid out $375,000 in a settlement involving the death of a patient because an infusion pump was erroneously set to deliver 210 cc of heparin per hour instead of the ordered 21 cc per hour [13].
- A fatal radiation overdose accident, involving a Therac radiation therapy device, occurred due to a human error [14].
- A patient died because of impeded airflow due to upside-down installation of an oxygen machine part [15].
- The FDA's Center for Devices and Radiological Health (CDRH) reported that around 60% of the deaths or serious injuries involving medical devices were the result of user errors [4].
- Over the years, many patient injuries and deaths occurred due to the insertion of a cassette by users from one infusion pump model into another incompatible model [15,16].
- A study of the FDA incident reports concerning Abbott Patient Controlled Analgesia (PCA) infuser reported that about 68% of the deaths and serious injuries were the result of human error [17].
- A group of medical device manufacturers carried out an investigation of design validation testing-related errors and discovered that 40% of the errors were due to coding or schematics implementation, 40% due to changes in requirements, 20% due to wrong test protocols or misinterpretation by the tester, and none due to design [18].

- A patient died while receiving oxygen when a concentrator pressure hose loosened and the intensity of alarm was too low to be heard properly over the drone of the device [15].

10.3 Medical Device Operator Errors and Medical Devices with a High Incidence of Human Errors

There are many operator-related errors that occur during the operation of medical devices or equipment. The important ones are shown in Figure 10.1 [19].

Over the years, various studies have been conducted to identify medical devices with a high incidence of human error. As the result of such studies, the most error prone medical devices (i.e., medical devices with a high

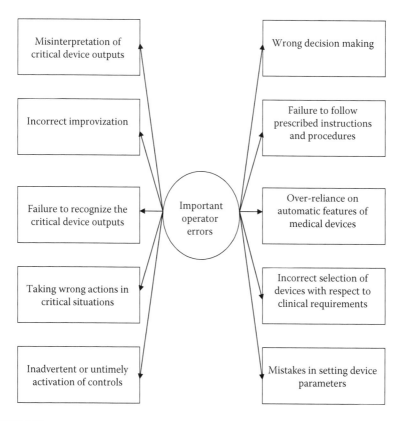

FIGURE 10.1
Important operator errors associated with medical devices or equipment.

incidence of human error) were identified [3,5,20]. Thus, the medical devices in the order of most error-prone to least error-prone are as follows [3,5,20]:

- Glucose meter
- Balloon catheter
- Orthodontic bracket aligner
- Administration kit for peritoneal dialysis
- Permanent pacemaker electrode
- Implantable spinal cord simulator
- Intravascular catheter
- Infusion pump
- Urological catheter
- Electrosurgical cutting and coagulation device
- Nonpowered suction apparatus
- Mechanical/hydraulic impotence device
- Implantable pacemaker
- Peritoneal dialysate delivery system
- Catheter introducer
- Catheter guide wire
- Transluminal coronary angioplasty catheter
- External low-energy defibrillator
- Continuous ventilators (respirators)

10.4 Human Error Causing User-Interface Device Design-Related Problems and Useful Guidelines for Medical Device Control/Display Arrangement and Design, Software Design, Installation, and Alarms with Respect to User Errors

Past experiences indicate that there are many user-interface device design-related problems that tend to "invite" the occurrence of user errors [21]. Some of these problems are presented below [21]:

- Inadequate or poorly designed labels
- Ambiguous or difficult to read displays

- Poor device design leading to unnecessarily complex maintenance and installation tasks
- Confusing or unnecessarily intrusive device associated alarms
- Complex or unconventional arrangements of items, such as controls, displays, and tubing
- Poor device feedback or status indication that lead to user uncertainty
- Difficult to remember, and/or confusing device operational instructions

Over the years, professionals working in the area of medical devices have developed various types of useful guidelines to reduce directly or indirectly the occurrence of user errors in medical devices. These guidelines for device control/display arrangement and design, software design, installation, and alarms are presented below [5,15].

10.4.1 Useful Guidelines for Device Control/Display Arrangement and Design to Reduce User Errors

Twelve of these guidelines include [15]:

1. Ensure that all switches and knobs are designed in such a way that they correspond to the conventions of all types of potential users.
2. Ensure that all device design facets are as consistent with potential users' expectations as possible.
3. Ensure that all control knobs, keys, and switches are spaced sufficiently apart for easy manipulation.
4. Ensure that all controls provide necessary tactile feedback.
5. Ensure that controls, displays, and workstations are designed on the basis of basic capabilities of the user (i.e., strength, reach, vision, dexterity, hearing, and memory).
6. Ensure that appropriate brightness of visual signals can be perceived by users working under varying illumination levels.
7. Ensure that all display and control arrangements are uncluttered and well organized.
8. Ensure that color and shape coding are used as necessary to facilitate the rapid identification of displays and controls.
9. Ensure that the intensity and pitch of auditory signals are able to be heard properly by device users.
10. Ensure the consistency of the acronyms, symbols, abbreviations, and text placed on the device by checking with the proper instructional manual.

11. Ensure that labels and displays are made in such a way that they can easily be read from typical viewing angles and distances.

12. Ensure that switches, keys, and knobs are arranged and designed so that they clearly minimize the likelihood of inadvertent activation.

10.4.2 Useful Guidelines for Device Software Design to Reduce User Errors

Nine of these guidelines include [15]:

1. Avoid contradicting the user expectations.

2. Avoid over-using software in situations where a simple hardware solution is clearly feasible.

3. Ensure that only those design procedures that entail easy-to-remember steps are used.

4. Ensure that the use and design of symbols, abbreviations, formats, and headings are unambiguous and consistent.

5. Ensure that only the accepted icons, abbreviations, colors, and symbols are used to convey information quickly, economically, and reliably.

6. Ensure that users are kept up to date regarding the current status of the device.

7. Ensure that all dedicated displays or display sectors for critical information are considered carefully.

8. Avoid confusing or overloading potential users with information that is too brief, densely packed, or unformatted.

9. Ensure that for correction and troubleshooting guides only conspicuous mechanisms are provided.

10.4.3 Useful Guidelines for Device Installation to Reduce User Errors

Six of these guidelines include [15]:

1. Ensure that cables, tubing, connectors, levers, and other hardware are designed for easy connection and installation.

2. Ensure that accessories and parts are numbered in such a way that the defective ones can be easily replaced with the good ones.

3. Avoid exposed electrical contacts.

4. Ensure that textual complexity in maintenance-related documents is lowered considerably by adding in the appropriate graphics.

5. Ensure that all the user instructions are easily comprehensible and the warnings are conspicuous.

6. Ensure that the positive locking mechanisms are present when there is a probability of compromising the connections integrity by factors such as part durability, motion, or casual contact.

10.4.4 Useful Guidelines for Device Alarms to Reduce User Errors

Eight of these guidelines include [15]:

1. Ensure that only those codes are used that correspond to established conventions.
2. Ensure that a wide spectrum of operating environments is considered in designing and testing alarms.
3. Ensure that proper priority is given to all critical alarms.
4. Ensure that both brightness contrast and color contrast are effective in varying illumination conditions.
5. Ensure that all auditory and visual alerts and critical alarms are clearly included in the device/equipment design specifications.
6. Ensure that alarms are designed in such a way that they satisfy or exceed normal visual and hearing limitations of typical users.
7. Ensure that all alarms are designed in such a way that they can easily be distinguished from one another, particularly from alarms on other medical devices within the area.
8. Ensure that proper consideration is given to the effects of static electricity, electromagnetic interference, and over-sensitivity on alarm operation.

10.5 General Guidelines for Reducing Medical Device/ Equipment User Interface-Related Errors

Past experiences indicate that medical device/equipment user interface design-related problems often lead to the occurrence of human errors. This section presents a number of general guidelines to make user interface designs of items, such as patient monitors, blood chemistry analyzers, infusion pumps, kidney dialysis machines, and ventilators, more user friendly, thus decreasing the probability of occurrence of user interface-related errors. These general guidelines are shown in Figure 10.2 [22].

The guideline "Simplify typography" calls for eliminating excessive highlighting, such as bolded, underlined, and italicized words. The guideline "Use simple and straightforward language" calls for avoiding the use of overly complex words and phrases. The guideline "Limit the number of

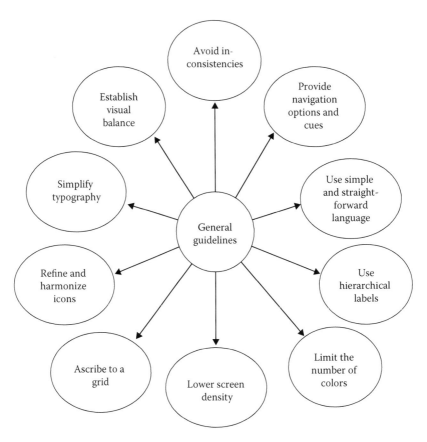

FIGURE 10.2
General guidelines for reducing medical device/equipment user-interface-related errors.

colors" calls for limiting the color palette of a user interface and keeping background and major on-screen components between three and five colors, including shades of grey. Also, choose colors with care so that they are consistent with medical conventions. The guideline "Lower screen density" calls for reducing the overstuffing of medical device/equipment displays with information because it could be difficult for medical professionals to pick out information at a glance from overly dense user interfaces.

The guideline "Provide navigation options and cues" calls for having appropriate navigation options and cues because moving from one place to another in a medical device/equipment user interface can sometimes cause users to become lost. The guideline "Ascribe to a grid" calls for fitting all on-screen elements into a grid because it eventually pays off in terms of perceived simplicity and visual appeal. Furthermore, as per past experiences, grid-based screens tend to be easy to implement in computer code because of the predictability of the position of visual elements.

The guideline "Establish visual balance" calls for creating visual balance or symmetry about the vertical axis. More specifically, arrange all visual elements on either side of an assumed axis in such a way that each side has roughly the same amount of content as empty space. In regard to the guideline "Use hierarchical labels," it is added that, as redundant labeling normally results in congested screens and takes a long time to scan, hierarchical labeling can help to save space and speed scanning by displaying items, such as heart rate, arterial blood pressure, and respiratory rate, more efficiently.

The guidelines "Avoid inconsistencies" and "Refine and harmonize icons" are considered self-explanatory, thus they are not described here.

10.6 An Approach to Human Factors during the Medical Device Development Process to Reduce Human Errors

Past experiences indicate that the occurrence of human errors in medical devices can be reduced quite significantly by making human factors an integral part of the medical devices' development process (i.e., from the concept phase to the production phase). The human factors aspects during the medical device development process phases are discussed below [23].

Concept phase. During this phase the human factors specialists works with marketing personnel, conducts interviews with potential device users, conducts analysis of regulatory and industry standards, helps to develop and implement appropriate questionnaires, and evaluates competitive medical devices. The specialist also examines with care the proposed operation of the device under consideration in regard to factors, such as skill range, experiences, and educational background of the intended users, and highlights the device's possible use environments.

Allocation of functions and preliminary design phase. During this phase, the human factors specialist and the design professionals determine which device functions will need manual points of interface between humans and the device and which will be automatic. More specifically, the manual points of interface are those functions where humans are required to monitor and control in such a way that the desired feedback or output from the device is obtained. Furthermore, the analysis of the preliminary design is conducted in regard to the operating environment of the device and the skill level of the most untrained users of the device. Normally, this task is carried out by considering the sketches or drawings of the expected operational environment and gauging reactions of expected users.

Preproduction prototype phase. During this phase the device prototype is built or updated for additional evaluation and market testing.

Market test and evaluation phase. This phase involves not only the actual testing of the medical device under consideration, but also a thorough evaluation of the feedback received from the market test by engineering, human factors, and marketing professionals.

Final design phase. During this phase, the design of the device is finalized by incorporating any human factors-related changes resulting from marketing, test, and evaluation.

Production phase. During this phase, the device is manufactured and put on the market. Also, during this phase, the human factors professional normally monitors the performance of the device, performs analysis of the proposed design-related changes, and provides assistance in the development of user training programs.

10.7 Causes and Classifications of Human Errors Resulting in Fatal Mine Accidents

Over the years there were various types of human errors that resulted in fatal mine accidents including directly or indirectly accidents associated with mining equipment. These errors occurred due to various causes that can be grouped under six classifications, as shown in Figure 10.3 [8,24–25].

Some of the main causes of classification "Failure to recognize a perceived warning" human error were poor inspection methods, neglect of proper instructions, obstruction of the line of sight, distraction or inattention, and masking noise. Similarly, some of the main causes of classification "Failure to perceive a warning" human error were inadequate information, lack of training, and lack of experience.

The causes of classifications "Ineffective secondary warning" and "Underestimation of a hazard," humans errors were unidentifiable. The main cause of classification "Failure to respond to a recognized warning" human error was underestimation of the hazard. Finally, three main causes of classification "Ineffective response to a warning" human error were inappropriate standard practice, carelessness or negligence, and well-intended but ineffective direct action.

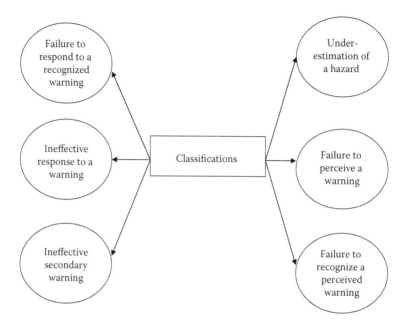

FIGURE 10.3
Classifications of fatal mine accidents' human error causes.

10.8 Common Mining Equipment-Related Maintenance Errors and Their Contributory Factors

There are many mining equipment-related maintenance errors. Some of the common ones include [24,26]:

- Failure to check, align, or calibrate
- Installation of wrong part
- Failure to detect while inspecting
- Parts installed backwards
- Failure to lubricate
- Error resulting from failure to complete task properly because of shift change
- Failure to follow prescribed instructions and procedures
- Use of wrong greases, fluids, or lubricants
- Failure to act on indicators of problems properly due to factors such as time constraints, priorities, or workload

- Omitting a part
- Reassemble error
- Failure to seal or close properly

Over the years, professionals working in the area of mining have identified many factors that contribute to the occurrence of mining equipment-related maintenance errors. Some of the main ones are [24,26]:

- Inaccessible parts or components
- Lack of proper tools and troubleshooting guides
- Poorly written manuals
- Inadequate task inspection and check-out time
- Poor layout of parts in a compartment
- Excessive weight of parts being manually handled
- Inability to make visual inspections
- Inappropriate placement of parts on equipment
- Confined work spaces
- Poor provision for cable and hose management

10.9 Useful Engineering Design Improvement Guidelines to Reduce Mining Equipment Maintenance Errors and General Factors Responsible for Failing to Reduce Human Errors in the Mining Sector at Large

Over the years engineering professionals working in the area of mining sector have developed many useful engineering design improvement guidelines to reduce the occurrence of mining equipment maintenance errors. Some of these guidelines include [26]:

- Improve items, such as warning readouts, devices, and indicators, to minimize or reduce human decision making.
- Improve equipment part interface by designing interfaces in such a way that the part can only be installed correctly and provide necessary mounting pins and other appropriate devices to support a part while it is being bolted or unbolted.
- Design to facilitate detection of human errors.
- Make use of decision guides to minimize or reduce human guesswork by providing arrows for indicating correct hydraulic pressures, flow direction, and correct type of lubricants or fluids.

- Use operational interlocks in such a way that subsystems cannot be turned on when they are incorrectly assembled or installed.
- Aim to improve fault isolation design by indicating the fault direction, providing built-in test capability, and designating appropriate test points and procedures.

There are many general factors that are responsible for failing to reduce human errors in the mining sector at large. Some of these factors are [27]:

- Nowadays mining workers are performing their assigned tasks under more difficult environmental, geo-mining, and physical conditions than ever before.
- Today's mining industry, because of a greater degree of automation and mechanization, requires greater efficiency, understanding, and capability from mining workers.
- Stress in the home is on the rise around the globe.
- Today's mining workers, in general, have greater mental tension and worry because of a greater desire to have more than others.

10.10 Methods to Perform Mining Equipment Human Error Analysis

There are many methods that can be used to conduct human error analysis of engineering systems or equipment [28]. Two of these methods considered most useful to perform human error analysis of mining equipment are presented below.

10.10.1 Probability Tree Method

This method is concerned with conducting task analysis by diagrammatically representing critical human actions and other events associated with the system under consideration. The branches of the probability tree denote diagrammatic task analysis. More specifically, the outcomes (i.e., success or failure) of each event are denoted by the tree branching limbs and, in turn, each branching limb is assigned an occurrence probability.

Four principal advantages of the probability tree method include [28]:

- A useful method for incorporating, with some modifications, factors, such as interaction stress, interaction effects, and emotional stress.
- A quite useful method to lower the occurrence probability of error due to computation resulting from computational simplification.

- A useful visibility tool.
- Λ quite useful method to readily estimate conditional probability, which may otherwise be obtained through rather complex probability equations.

The application of the method to perform mining equipment human error analysis is demonstrated through the following example.

Example 10.1

Assume that a maintenance worker performs three consecutive tasks x, y, and z concerned with a piece of mining equipment and each of these tasks can either be performed correctly or incorrectly. The incorrect performance of any of these three tasks can lead to a serious accident. The task x is performed before task y and in turn task y is carried out before task z, and all of these three tasks are independent of each other.

Develop a probability tree and obtain a probability expression that the mining equipment maintenance worker will fail to accomplish the overall mission successfully.

A probability tree for this example is shown in Figure 10.4. The tree shows that the mining equipment maintenance worker first performs task x correctly or incorrectly and then proceeds to task y, which also can be performed correctly or incorrectly, and then finally the worker proceeds to task z, which also can be performed correctly or incorrectly.

It is to be noted that in Figure 10.4, single small letters x, y, and z without bars denote successful events (i.e., tasks x, y, and z performed correctly) and with bars denote unsuccessful events (i.e., tasks, x, y, and z performed incorrectly). Other symbols used to obtain the probability expression for the example are defined below, where:

P_x = the probability of performing task x correctly.
P_y = the probability of performing task y correctly.
P_z = the probability of performing task z correctly.
$P_{\bar{x}}$ = the probability of performing task x incorrectly.
$P_{\bar{y}}$ = the probability of performing task y incorrectly.
$P_{\bar{z}}$ = the probability of performing task z incorrectly.
P_{mwf} = the probability of failure of the maintenance worker to accomplish the overall mission successfully.

By using the Figure 10.4 diagram and the above symbols, we write down the following probability expression for the maintenance worker not accomplishing the overall mission successfully:

$$P_{mwf} = P_x P_y P_{\bar{z}} + P_{\bar{x}} P_y P_z + P_{\bar{x}} P_y P_{\bar{z}} + P_{\bar{x}} P_{\bar{y}} P_z$$
$$+ P_x P_{\bar{y}} P_z + P_x P_{\bar{y}} P_{\bar{z}} + P_{\bar{x}} P_y P_{\bar{z}}$$

(10.1)

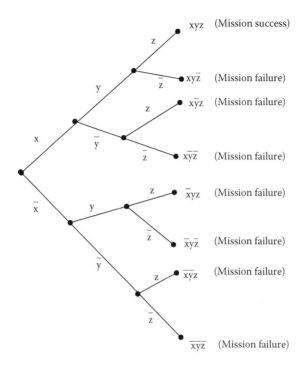

FIGURE 10.4
Probability tree for Example 10.1.

Thus, the expression for the probability that the mining equipment maintenance worker will fail to accomplish the overall mission successfully is given by equation (10.1).

10.10.2 Fault Tree Analysis

This method is widely used in the industrial sector to perform various types of reliability and safety analyses and it also can be used to conduct mining equipment human error analysis. The method is described in Chapter 4 and in Refs. [29,30]. The following example demonstrates the application of the method to conduct mining equipment human error analysis.

Example 10.2

Assume that a mining equipment maintenance worker can commit an error due to five factors and the error can result in a serious accident. The five factors are carelessness, inadequate tools, poor environment, poor instructions, or inadequate training. In turn, two principle reasons for the poor instructions are poor verbal instructions or poorly written maintenance procedures.

Similarly, two principal reasons for the poor environment are poor access or too much noise.

Develop a fault tree for the top event "Mining equipment maintenance worker committed an error" by using the fault tree symbols given in Chapter 4, and calculate the top event occurrence probability if the probabilities of occurrence of independent events carelessness, inadequate tools, inadequate training, poor verbal instructions, poorly written maintenance procedures, poor access, and too much noise are 0.01, 0.02, 0.05, 0.06, 0.07, 0.08, and 0.09, respectively.

A fault tree for the example is shown in Figure 10.5. The single capital letters in the fault tree diagram denote corresponding event (e.g., H for poorly written maintenance procedures and A for poor environment).

Using an equation given in Chapter 4 for the probability of occurrence of the OR gate output fault event and the specified data values, the probability of occurrence of event A in Figure 10.5 is

$$P(A) = 1 - (1 - P(F))(1 - P(G))$$

$$= 1 - (1 - 0.08)(1 - 0.09)$$

$$= 0.1628$$

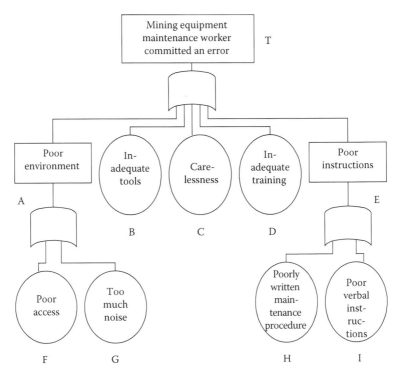

FIGURE 10.5
A fault tree for Example 10.2.

where
P(F) = the probability of occurrence of event F.
P(G) = the probability of occurrence of event G.

Similarly, using the same equation given in Chapter 4 for the probability of occurrence of the OR gate output fault event and the given data value, the probability of occurrence of event E in Figure 10.5 is

$$P(E) = 1 - (1 - P(H))\,(1 - P(I))$$

$$= 1 - (1 - 0.07)\,(1 - 0.06)$$

$$= 0.1258$$

where
P(H) = the probability of occurrence of event H.
P(I) = the probability of occurrence of event I.

Using the above calculated values, the specified data values, and the equation given in Chapter 4 for the probability of occurrence of the OR gate output fault event, the probability of occurrence of top event T in Figure 10.5 is

$$P(T) = 1 - (1 - P(A))\,(1 - P(B))\,(1 - P(C))\,(1 - P(D))\,(1 - P(E))$$

$$= 1 - (1 - 0.1628)(1 - 0.02)(1 - 0.01)(1 - 0.05)(1 - 0.1258)$$

$$= 0.3254$$

Thus, the probability of the occurrence of the top event T (i.e., mining equipment maintenance worker committed an error) is 0.3254.

Problems

1. Write an essay on human error in healthcare systems and in mining equipment.
2. List at least eight facts and figures concerned with healthcare systems.
3. Discuss medical device operator errors.
4. List at least 10 medical devices with a high incidence of human errors.
5. What are the human errors causing user-interface medical device design-related problems?

6. Describe the approach to human factors during the medical device development process to reduce human errors.
7. What are the main causes of human errors that lead to fatal mine accidents?
8. What are the common mining equipment-related maintenance errors and their contributory factors?
9. List and discuss at least five useful engineering design improvement guidelines to reduce mining equipment maintenance errors.
10. List at least five useful guidelines for medical device installation to reduce user errors.

References

1. Murray, K., Canada's Medical Device Industry Faces Cost Pressures, Regulatory Reform, *Medical Device and Diagnostic Industry Magazine*, Vol. 19, 1997, pp. 30–39.
2. Hutt, P. B., A History of Government Regulation of Adulteration and Misbranding of Medical Devices, in *The Medical Device Industry*, ed. N. F. Estrin, Marcel Dekker, New York, 1990, pp. 17–33.
3. Dhillon, B. S., *Medical Device Reliability and Associated Areas*, CRC Press, Boca Raton, FL, 2000.
4. Bogner, M. S., Medical Devices: A New Frontier for Human Factors, *CSERIAC Gateway*, Vol. IV, No. 1, 1993, pp. 12–14.
5. Dhillon, B. S., *Human Reliability and Error in Medical System*, World Scientific Publishing, River Edge, NJ, 2003.
6. Dhillon, B. S., *Reliability Technology, Human Error, and Quality in Health Care*, CRC Press, Boca Raton, FL, 2008.
7. Chadwick, J., Higgins, S., US Technology, *International Mining*, September 2006, pp. 44–54.
8. Lawrence, A. C., Human Error as a Cause of Accidents in Gold Mining, *Journal of Safety Research*, Vol. 6, No. 2, 1974, pp. 78–88.
9. Maddox, M. E., Designing Medical Devices to Minimize Human Error, *Medical Device and Diagnostic Industry Magazine*, Vol. 19, No. 5, 1997, pp. 160–180.
10. Bogner, M. S., Medical Devices and Human Error, in *Human Performance in Automated Systems: Current Research and Trends*, eds. M. Moulona, R. Parasuraman, Lawrence Erlbaum Associates Publishers, Hillsdale, NJ, 1994, pp. 64–67.
11. Nobel, J. L., Medical Device Failures and Adverse Effects, *Pediatric Emergency Care*, Vol. 7, 1991, pp. 120–123.
12. Improving Patient Care by Reporting Problems with Medical Devices, A MedWatch Continuing Education Article, Food and Drug Administration (FDA), Rockville, MD, September 1997, pp. 1–8.
13. Brueley, M. E., *Ergonomic and Error: Who is Responsible?* Proceedings of the First Symposium on Human Factors in Medical Devices, 1989, pp. 6–10.

14. Casey, S., *Set Phasers on Stun and Other True Tales of Design Technology and Human Error*, Aegean Inc., Santa Barbara, CA, 1993.
15. Swayer, D., *Do It by Design: An Introduction to Human Factors in Medical Devices*, Center for Devices and Radiological Health, Food and Drug Administration, Washington, D.C., 1996.
16. Kortstra, J. R. A., Designing for the User, *Medical Device Technology*, January/February, 1995, pp. 22–28.
17. Lin, L., *An Ergonomically Redesigned Analgesia Delivery Device Proves Safer and More Efficient*, Proceedings of the Human Factors and Ergonomics Society Annual Meeting, 1998, pp. 346–350.
18. Olivier, D. P., Engineering Process Improvement through Error Analysis, *Medical Device and Diagnostic Industry Magazine*, Vol. 21, No. 3, 1999, pp. 194–202.
19. Hyman, W. A., Human Factors in Medical Devices, in *Encyclopedia of Medical Devices and Instrumentation,* ed. J. G. Webster, Vol. 3, John Wiley & Sons, New York, 1988, p. 1542–1553.
20. Wiklund, M. E., *Medical Device and Equipment Device*, Interpharm Press, Inc., Buffalo Grove, IL, 1995.
21. Rachlin, J. A., Human Factors and Medical Devices, *FDA User Facility Reporting*: *A Quarterly Bulletin to Assist Hospitals, Nursing Homes, and Other Device User Facilities*, No. 12, 1995, pp. 86–89.
22. Wiklund, M. E., Making Medical Device Interfaces More User-Friendly, *Medical Device and Diagnostic Industry Magazine*, Vol. 20, No. 5, 1998, pp. 177–186.
23. Le Cocq, A. D., Application of Human Factors Engineering in Medical Product Design, *Journal of Clinical Engineering*, Vol. 12, No. 4, 1987, pp. 271–277.
24. Dhillon, B. S., *Mining Equipment Reliability, Maintainability, and Safety,* Springer, Inc., London, 2008.
25. Dhillon, B. S., *Mine Safety: A Modern Approach*, Springer, Inc., London, 2009.
26. Under, R. L., Conway, K., Impact of Maintainability Design on Injury Rates and Maintenance Costs for Underground Mining Equipment, in *Improving Safety at Small Underground Mines*, compiled by R. H. Peters, Special Publication No. 18-94, Bureau of Mines, U.S. Department of the Interior, Washington, D.C., 1994.
27. Mohan, S., Duarte, D., *Cognitive Modeling of Underground Miners Response to Accidents,* Proceedings of the Annual Reliability and Maintainability Symposium, 2006, pp. 51–55.
28. Dhillon, B. S., *Human Reliability: With Human Factors*, Pergamon Press, New York, 1986.
29. Dhillon, B. S., Singh, C., *Engineering Reliability: New Techniques and Applications*, John Wiley & Sons, New York, 1981.
30. Dhillon, B. S., *Design Reliability: Fundamentals and Applications*, CRC Press, Boca Raton, FL, 1999.

11

Human Error in Power Plant Maintenance and Aviation Maintenance

11.1 Introduction

Although, over the years, impressive progress has been made in maintaining engineering equipment in the field environment, maintenance of equipment or systems is still a challenging issue due to factors such as cost, complexity, size, and competition. Each year billions of dollars are spent on engineering equipment maintenance around the globe. For example, as per Ref. [1], each year in the United States alone, over $300 billion is spent on plant maintenance by the U.S. industrial sector and about 80% of this amount is spent to rectify the chronic failure of systems, machines, and people.

Maintenance is an essential activity in power plants and, over the years, it has been found that human error in maintenance is an important factor in the causation of power generation safety-related incidents [2]. For example, a study of reliability problem-related events concerning electrical/electronic parts in nuclear power generation plants reported that human errors made by maintenance personnel exceeded operator errors [2,3].

Maintenance is an important element of the aviation industry. For example, in 1989, the United States airlines spent approximately 12% of operating costs on the maintenance activity and during the period 1980 to 1988, the airlines maintenance cost jumped from around $2.9 billion to $5.7 billion [4–6]. Human error in aviation maintenance is increasingly becoming an important issue worldwide. For example, a study performed in the United Kingdom revealed that the maintenance error occurrence events per million flights has doubled from 1990 to 2000 [7].

This chapter presents various important aspects of human error in power plant maintenance and aviation maintenance.

11.2 Power Plant Maintenance Human Error-Related Facts, Figures, and Examples

Some of the power plant maintenance human error-related facts, figures, and examples are:

- A number of studies conducted over the years reported that roughly between 55 and 65% of human performance-related problems surveyed in the area of power generation were concerned with maintenance activities [8,9].
- A study of 199 human errors that occurred in Japanese nuclear power plants during the period from 1965 to 1995 reported that approximately 50 of them were concerned with maintenance activities [10].
- A blast at the Ford Rouge power generation plant in Dearborn, Michigan, that caused six deaths and many injuries was the result of a maintenance error [11,12].
- A study revealed that over 20% of all types of system failures in fossil power plants occur due to various types of human errors, and maintenance errors account for roughly 60% of the annual power loss due to human-related errors [13].
- A study of 126 human error-associated events in 1990, in the area of nuclear power generation, revealed that around 42% of the problems were directly or indirectly linked to maintenance and modification [8].
- In 1989 on Christmas day, two nuclear reactors in the state of Florida were shut down due to a maintenance error and caused rolling blackouts [14].
- A study of around 4400 maintenance history records covering the period from 1992 to 1994, concerning a boiling water reactor nuclear power plant, revealed that approximately 7.5% of all failure records could be categorized as human errors related to maintenance activities [15,16].

11.3 Classifications of Causes for the Occurrence of Human Errors in Power Plant Maintenance and Their Causal Factors

There are many causes for the occurrence of human errors in power plant maintenance. On the basis of characteristics obtained from modeling the maintenance task, these causes may be classified under the following four classifications [2,17]:

Classification I: Design shortcomings in hardware and software. This classification includes items, such as deficiencies in the design of displays and controls, incorrect or confusing procedures, and insufficient communication equipment.

Classification II: Disturbances of the external environment. Some examples of disturbances of the external environment are the physical conditions, such as temperature, ventilation, ambient illumination, and humidity.

Classification III: Human ability limitations. An important example of human ability limitations is the limited capacity of an individual's short-term memory in the internal control mechanism.

Classification IV: Induced circumstances. This classification includes items such as emergency conditions, momentary distractions, and improper communications that may result in failures.

There are many causal factors for the occurrence of human errors in power plant maintenance. A study identified causal factors presented in Table 11.1, in order of greatest to least frequency of occurrence, for critical incidents and reported events concerning maintenance error in power generation plants [18,19].

Faulty procedures are the most frequent causal factor in the reported mishaps. It includes items such as lack of adherence to an outlined procedure, wrong procedures, lack of specificity, and incompleteness. A good example of faulty procedures is "because of poor judgment and not following stated guidelines properly, a ground was left on a circuit breaker. When the system or equipment was put back into operation, the circuit breaker in

TABLE 11.1

Causal Factors in Order of Greatest to Least Frequency of Occurrence, for Critical Incidents and Reported Events Concerning Maintenance Error in Power Generation Plants

No.	Causal Factor
1	Faulty procedure
2	Problems in clearing and tagging equipment for maintenance
3	Shortcomings in equipment design
4	Problems in moving people or equipment
5	Poor training
6	Poor unit and equipment identification
7	Problems in facility design
8	Poor work practices
9	Adverse environmental factors
10	Mistakes by maintenance personnel

question blew up and caused quite high property damage." In this situation, the proper or correct procedure would have required clearing the ground before putting back the circuit breaker into operation.

The second most frequent causal factor in reported cases—Problems in clearing and tagging equipment for maintenance—is where serious accidents or potentially serious accidents could be attributed to an error or failure related to the equipment clearance process. The third most frequent causal factor for near-accidents/accidents—Shortcomings in equipment design—revolved around equipment design-associated problems. Some of the items included in the factor include:

- The equipment not designed with appropriate mechanical safeguards for preventing the substitution of incorrect part for the proper replacement part.
- Poorly designed and inherently unreliable parts.
- Parts placed in inaccessible locations.
- Equipment installed incorrectly right from the outset.

The fourth most frequent causal factor—Problems in moving people or equipment—are the problems that basically stem from the inability to use proper vehicular aids to move heavy units of equipment or poor lifting capability. The fifth most frequent causal factors are Poor training, Poor unit and equipment identification, and Problems in facility design. Poor training, the fifth most frequent causal factor, is basically concerned with the unfamiliarity of repair personnel with the job or their lack of proper awareness of the system characteristics and all types of inherent dangers associated with the job at hand.

The factor, Poor unit and equipment identification, is the cause of a high number of accidents, and frequently the problem is confusion between identical items and sometimes unclear identification of potential hazards. The factor, Problems in facility design, can contribute to various types of accidents. Two typical examples of these problems are improperly sized facilities causing an overly dense packaging of equipment systems and preventing proper performance of inspection or repair tasks and inadequate clearances for repair personnel, equipment, or transportation aids in the performance of maintenance tasks.

The sixth most frequent causal factor is Poor work practices. Two examples of poor work practices are not taking the time to erect a proper scaffold so that an item in midair can be accessed safely and not waiting for operators to finish the tagging and switching tasks required to disable the systems requiring attention.

Finally, the last two most frequent causal factors are Adverse environmental factors and Mistakes by maintenance personnel. The factor, Adverse environmental factors, includes items such as the necessity to wear appropriate devices and garments in threatening environments that, in turn, restrict visual field and movement capabilities of an individual, and the encouragement of haste by the necessity to reduce stay time in unfriendly environments. The factor, Mistakes by maintenance personnel, is a small fraction of those types of errors that would be rather difficult to anticipate and "design out" of power plants altogether.

11.4 Maintenance-Related Tasks Most Susceptible to the Occurrence of Human Error in Power Generation

In the 1990s, a joint study by the Electric Power Research Institute (EPRI) in the United States and Central Research Institute of Electric Power Industry (CRIEPI) in Japan was performed to highlight critical maintenance-related tasks and to develop, implement, and evaluate appropriate interventions having a high potential to decrease human errors or increasing maintenance-related productivity in nuclear power generation plants. As the result of this study, the following five maintenance tasks considered most susceptible to the occurrence of human errors were identified [20]:

1. Replace reactor coolant pump (RCP) seals
2. Test reactor protection system (RPS)
3. Overhaul mainstream isolation valves (MSIV)
4. Overhaul motor operated valve (MOV) actuator
5. Overhaul main feed water pump (MFWP)

It means that the above maintenance tasks must be performed with utmost care in order to minimize or eliminate the occurrence of human errors.

11.5 Methods for Performing Human Error Analysis in Power Plant Maintenance

There are many methods that can be used to perform human error analysis in power plant maintenance. Two of these methods are described below.

11.5.1 Maintenance Personnel Performance Simulation (MAPPS) Model

This model/method was developed by the Oak Ridge National Laboratory to provide estimates of nuclear power plant (NPP) maintenance manpower performance measures [21]. The main objective for the development of this model or method was the definite need for and lack of a human reliability data bank specifically pertaining to NPP maintenance activities for use in conducting probabilistic risk assessment-related studies.

The main measures of performance estimated by this model include:

- Probability of an undetected error
- The task duration time
- Identification of the most- and least-likely error-prone subelements
- Maintenance team stress profiles during the task execution process
- Probability of successfully completing the task of interest

All in all, the MAPPS model is a quite powerful tool to estimate various types of important maintenance parameters, and its flexibility permits it to be useful for types of applications concerning NPP maintenance activity. Additional information on this method is available in Ref. [21].

11.5.2 Markov Method

This is a commonly used method to conduct various types of reliability-related analysis of engineering systems, and it also can be used to conduct human error analysis in the area of power plant maintenance. The method is described in Chapter 4. Its application to conduct human error analysis in power plant maintenance is demonstrated through the following example.

Example 11.1

Assume that equipment used in a power plant can fail due to a maintenance error or nonmaintenance error problems or failures. The equipment state space diagram is shown in Figure 11.1. Numerals in circles and box denote equipment states. The equipment maintenance error rate, nomaintenance error failure rate, and repair rates are constant. In addition,

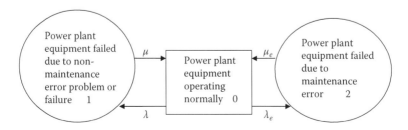

FIGURE 11.1
Power plant equipment state space diagram.

the repaired equipment is as good as new and equipment failures occur independently.

Develop expressions for the power plant equipment time dependent state probabilities, steady state probabilities, and time dependent and steady state availabilities and unavailabilities by using the Markov method.

The following symbols are associated with the Figure 11.1 diagram:

J = the power plant equipment state j; for j = 0 (power plant equipment operating normally), j = 1 (power plant equipment failed due to non-maintenance error problem or failure), j = 2 (power plant equipment failed due to maintenance error).

λ_e = the power plant equipment constant maintenance error rate.

λ = the power plant equipment constant nonmaintenance error problem or failure rate.

μ_e = the power plant equipment constant repair rate from state 2 to state 0.

μ = the power plant equipment constant repair rate from state 1 to state 0.

$P_j(t)$ = the probability that the power plant equipment is in state j at time t; for j = 0, 1, 2.

By using the Markov method described in Chapter 4, we write down the following equations for Figure 11.1:

$$\frac{dP_0(t)}{dt} + (\lambda + \lambda_e)P_0(t) = \mu P_1(t) + \mu_e P_2(t) \tag{11.1}$$

$$\frac{dP_1(t)}{dt} + \mu P_1(t) = \lambda P_0(t) \tag{11.2}$$

$$\frac{dP_2(t)}{dt} + \mu_e P_2(t) = \lambda_e P_0(t) \tag{11.3}$$

At time $t = 0$, $P_0(0) = 1$, $P_1(0) = 0$, and $P_2(0) = 0$.
By solving equation (11.1) to equation (11.3), we obtain

$$P_0(t) = \frac{\mu\mu_e}{y_1 y_2} + \left[\frac{(y_1 + \mu_e)(y_2 + \mu)}{y_1(y_1 - y_2)}\right]e^{y_1 t} - \left[\frac{(y_2 + \mu_e)(y_2 + \mu)}{y_2(y_1 - y_2)}\right]e^{y_2 t} \tag{11.4}$$

where

$$y_1, y_2 = \frac{-b \pm \sqrt{b^2 - 4(\mu\mu_e + \lambda_e\mu + \lambda\mu_e)}}{2} \tag{11.5}$$

$$b = \mu + \mu_e + \lambda + \lambda_e \tag{11.6}$$

$$y_1 y_2 = \mu\mu_e + \lambda_e\mu + \lambda\mu_e \tag{11.7}$$

$$y_1 + y_2 = -(\mu_e + \mu + \lambda_e + \lambda) \tag{11.8}$$

$$P_1(t) = \frac{\lambda_e\mu}{y_1 y_2} + \left[\frac{\lambda y_1 + \lambda\mu_e}{y_1(y_1 - y_2)} \right] e^{y_1 t} - \left[\frac{(\mu_e + y_2)\lambda_e}{y_2(y_1 - y_2)} \right] e^{y_2 t} \tag{11.9}$$

$$P_2(t) = \frac{\lambda\mu_e}{y_1 y_2} + \left[\frac{\lambda y_1 + \lambda\mu_e}{y_1(y_1 - y_2)} \right] e^{y_2 t} - \left[\frac{(\mu_e + y_2)\lambda}{y_2(y_1 - y_2)} \right] e^{y_2 t} \tag{11.10}$$

As time t becomes very large, we get the following steady-state probability equations from equation (11.4), equation (11.9), and equation (11.10), respectively:

$$P_0 = \frac{\mu\mu_e}{y_1 y_2} \tag{11.11}$$

$$P_1 = \frac{\lambda_e\mu}{y_1 y_2} \tag{11.12}$$

$$P_2 = \frac{\lambda\mu_e}{y_1 y_2} \tag{11.13}$$

where
P_0, P_1 and P_2 = the steady-state probabilities of the power plant equipment being in states 0, 1, and 2, respectively.

The power plant equipment time-dependent availability and unavailability are given by equation (11.14) and equation (11.15).

$$AV_{pe}(t) = P_0(t) \tag{11.14}$$

where
$AV_{pe}(t)$ = the power plant equipment availability at time t.

$$UAV_{pe} = P_1(t) + P_2(t) \tag{11.15}$$

where
$UAV_{pe}(t)$ = the power plant equipment unavailability at time t.

Similarly, the power plant equipment steady-state availability and unavailability are given by equation (11.16) and equation (11.17), respectively.

$$AV_{pe} = P_0 = \frac{\mu\mu_e}{y_1 y_2}$$ (11.16)

where
AV_{pe} = the power plant equipment steady state availability.

$$UAV_{pe} = P_1 + P_2 = \frac{\lambda_e \mu}{y_1 y_2} + \frac{\lambda \mu_e}{y_1 y_2}$$ (11.17)

where
UAV_{pe} = the power plant equipment steady-state unavailability.

Example 11.2

Assume that we have the following values for a power plant equipment:

$\lambda = 0.005$ failures per hour
$\mu = 0.03$ repairs per hour
$\lambda_e = 0.002$ errors per hour
$\mu_e = 0.01$ repairs per hour

Calculate the power plant equipment steady-state unavailability.
By substituting the given data values into equation (11.17), we get:

$$UAV_{pe} = \frac{(0.002)(0.03) + (0.005)(0.01)}{(0.03)(0.01) + (0.002)(0.03) + (0.005)(0.01)}$$

$$= 0.2683$$

Thus, the power plant equipment steady-state unavailability is 0.2683.

11.6 Guidelines to Reduce and Prevent Human Errors in Power Generation Maintenance

Over the years, professionals working in the area of power generation maintenance have proposed various guidelines to reduce and prevent human errors in power generation maintenance. Four of these guidelines

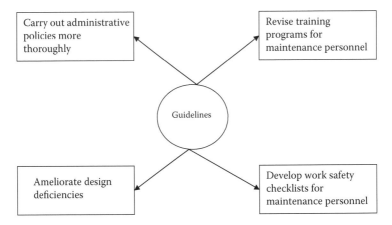

FIGURE 11.2
Guidelines to reduce and prevent human errors in power generation maintenance.

considered quite useful are shown in Figure 11.2 [2]. The guideline, Revise training programs for maintenance personnel, basically calls for revising training programs for maintenance personnel according to the characteristics and frequency of occurrence of each extrinsic cause. The guideline, Carry out administrative policies more thoroughly, basically calls for motivating maintenance personnel to comply with prescribed quality control procedures.

The guideline, Ameliorate design deficiencies, calls for overcoming deficiencies in areas such as coding, plant layout, work environment, and labeling as deficiencies in design can reduce attention to the tasks and may even induce human error. Finally, the guideline, Develop work safety checklists for maintenance personnel, calls for providing maintenance personnel with work safety checklists that can be utilized to determine the possibility of human error occurrence and the factors that may affect their actions before or after the performance of maintenance tasks.

11.7 Aviation Maintenance Human Error-Related Facts and Figures

Some of the aviation maintenance human error-related facts and figures include:

- Maintenance-related errors contribute to about 15% of air carrier accidents and cost over $1 billion dollars annually to the U.S. industry [22].

- For the period 1982 to 1991, an analysis of safety issues versus onboard fatalities among jet fleets worldwide highlighted maintenance and inspection as the second most important safety issue with onboard fatalities [23,24].
- In 1991, a human error during scheduled maintenance caused 13 fatalities in an Embraer 120 aircraft accident [7,25].
- A study of 122 maintenance-related errors occurring in a major airline over a three-year period reported that their breakdowns were: omission (56%), incorrect installations (30%), wrong parts (8%), and other (6%) [26,27].
- A Boeing study reported that 19.1% of in-flight engine shutdowns are due to maintenance error [22].
- In 1979, a DC-10 aircraft accident due to improper maintenance procedures followed by maintenance personnel caused 272 fatalities [28].

11.8 Causes for the Occurrence of Human Error in Aviation Maintenance

There are a large number of factors that can directly or indirectly impact aviation maintenance personnel performance. A document prepared by the International Civil Aviation Organization (ICAO) lists over 300 such factors/influences ranging from boredom to temperature [29]. Nonetheless, some of the important causes for the occurrence of human error in aviation maintenance include [28,30]:

- Poor equipment design
- Poor work layout
- Time pressure
- Inadequate training, work tools, and experience
- Poor work environment (e.g., temperature, lighting, and humidity)
- Outdated maintenance manuals
- Poorly written maintenance procedures
- Complex maintenance tasks
- Fatigued maintenance personnel

Additional information on the causes for the occurrence of human error in aviation maintenance is available in Ref. [29].

11.9 Types of Human Errors in Aircraft Maintenance Activities and Their Occurrence Frequency and Common Human Errors in Aircraft Maintenance

A Boeing study in 1994 examined 86 reports concerning aircraft incidents in regard to maintenance error and reported 31 types of human errors in aircraft maintenance activities. These types, along with their occurrence frequency in parentheses are as follows [31]:

- Incorrect part/equipment installed (1)
- Incorrect panel installation (1)
- Incorrect orientation (1)
- Incorrect fluid type (1)
- Necessary servicing not performed (1)
- Access panel not closed (1)
- Equipment not installed (1)
- Material left in aircraft/ engine (1)
- Unable to access part in stores (1)
- Contamination of open system (1)
- Warning sign or tag not used (2)
- Safety lock or warning moved (2)
- Vehicle/equipment made contact with aircraft (2)
- Vehicle driving instead of towing (2)
- Pin/tie left in place (2)
- Not tested properly (2)
- No proper verbal warning given (3)
- Unserviceable equipment used (4)
- Did not use or obtain proper equipment (4)
- Equipment not activated/ deactivated (4)
- Person came in contact with hazard (4)
- Person entered dangerous zones (5)
- Work not documented (5)
- Unfinished installation (5)
- Degradation not discovered (6)
- Falls and spontaneous actions (6)
- Equipment failure (10)
- System not made safe (10)
- Towing event (10)
- System operated in unsafe conditions (16)
- Miscellaneous (6)

Over the years, various studies have been performed to identify commonly occurring human errors in aircraft maintenance. One of these studies performed by the U.K. Civilian Aviation Authority (UKCAA) over a three-year period has identified the following eight commonly occurring human errors in aircraft maintenance [27,32]:

- Inadequate lubrication
- Loose objects, such as tools, left in the aircraft

- Wrong installation of parts
- Fitting of incorrect parts
- Failure to remove landing gear ground lock pins prior to aircraft departure
- Discrepancies in electrical wiring including cross connections
- Unsecured fairings, cowlings, and access panels
- Unsecured fuel caps and refuel panels

11.10 Maintenance Error Decision Aid (MEDA)

Maintenance error decision aid is an important tool to investigate contributing factors for the occurrence of human errors in the area of aviation in regard to maintenance. MEDA was developed in the 1990s by Boeing, along with industry partners, such as United Airlines and Continental Airlines, and it may simply be described as a structured process to investigate human error causes by aircraft maintenance personnel [33-35]. The philosophy of the process is as follows [35]:

- Human errors in aircraft maintenance result from a series of contributing factors.
- Many of the error contributing factors are part of airline processes and can be managed.
- Maintenance personnel do not make errors intentionally.
- Some maintenance error will not have specific corrective measures.

Thus, the four main objectives of the MEDA are [35]:

Objective I: To provide a proper means of human error trend analysis for the aircraft maintenance organization.

Objective II: To identify aircraft maintenance system-associated difficulties that increase exposure to human error and decrease efficiency.

Objective III: To provide the line-level aircraft maintenance manpower a standardized mechanism for investigating the occurrence of maintenance errors.

Objective IV: To provide the aircraft maintenance organization a better understanding of how human performance-related issues contribute to human error occurrence.

11.11 Guidelines to Reducing Human Error in Aircraft Maintenance

There are many useful guidelines to reduce the occurrence of human error in aircraft maintenance. They cover many areas, as shown in Figure 11.3 [17,24,36].

Some of the important guidelines in the area of human error risk management include:

- Avoiding performing simultaneously the same maintenance task on similar redundant units.
- Reviewing the need to disturb the normal operating systems to perform rather nonessential periodic maintenance because the disturbance may lead to a maintenance error.
- Reviewing formally the effectiveness of defenses, such as engine runs, built into the system to detect maintenance errors.

Two important guidelines in the area of design are:

- Ensuring that equipment manufacturers give appropriate attention to human factors concerning maintenance during the design phase.

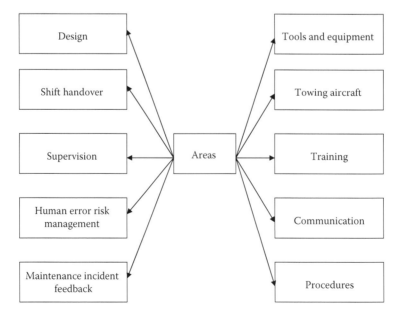

FIGURE 11.3

Areas covered by guidelines to reduce human error in aircraft maintenance.

- Actively seeking all types of relevant information on the occurrence of human errors during maintenance activities, to provide effective inputs in the design phase.

In the area of training, two particular guidelines are:

- Considering introducing crew resourcement for maintenance personnel.
- Providing training courses to all personnel involved with the maintenance activity with emphasis on company procedures.

In the area of communication, one useful guideline is to ensure that appropriate systems are in place for disseminating important information to all maintenance personnel so that changing procedures or repeated errors are considered with utmost care. Two important guidelines concerning the maintenance incident feedback area include:

- Ensuring that management personnel are given proper feedback on the occurrence of human factors-associated maintenance incidents on a regular basis, with appropriate consideration to the situations that play a pivotal role in the occurrence of such incidents.
- Ensuring that all personnel involved with the training activity are provided appropriate feedback on the occurrence of human factors-associated maintenance incidents on a regular basis, so that proper corrective measures aimed at these problems are taken in an effective manner.

Two particular guidelines concerning the area of tools and equipment are:

- Ensuring the storage of lockout devices in such a way that it becomes immediately apparent when they are misplaced inadvertently.
- Reviewing systems by which items, such as lighting systems and stands, are kept to remove unserviceable equipment from service and repairing it as fast as possible.

Three important guidelines in the area of procedures include:

- Reviewing maintenance work practices on a regular basis to ensure that they do not vary significantly from formal procedures.
- Ensuring that standard work practices are being followed properly throughout aircraft maintenance operations.
- Reviewing all documented maintenance procedures and practices on a regular basis in regard to items, such as consistency, realism, and accessibility.

One important guideline in the area of shift handover is to ensure the effectiveness of practices associated with this handover by carefully considering factors, such as documentation and communication, so that unfinished tasks are transferred properly and correctly across all shifts.

Similarly, one particular guideline pertaining to the area of supervision is to recognize that supervision- and management-associated oversights must be strengthened, especially in the final hours of all shifts because the occurrence of human errors becomes more likely at this time. Finally, one important guideline in the area of towing aircraft is to review periodically with care the equipment and procedures used for towing to and from maintenance facilities.

Problems

1. List at least five power plant maintenance human error-related facts and figures.
2. What are the maintenance-related tasks most susceptible to the occurrence of human error in power generation?
3. Describe maintenance personnel performance simulation (MPPS) model.
4. Prove equation (11.11) through equation (11.13) by using equation (11.4), equation (11.9), and equation (11.10).
5. What are the useful guidelines to reduce and prevent the occurrence of human errors in power generation maintenance?
6. List at least five aviation maintenance human error-related facts and figures.
7. List and discuss at least seven causes for the occurrence of human error in aviation maintenance.
8. What are the eight commonly occurring human errors in aircraft maintenance?
9. Describe maintenance error decision aid (MEDA).
10. Prove equation (11.4).

References

1. Latino, C. J., *Hidden Treasure: Eliminating Chronic Failures Can Cut Maintenance Costs Up to 60%*, Report, Reliability Center, Hopewell, Virginia, 1999.

2. Wu, T. M., Hwang, S. L., Maintenance Error Reduction Strategies in Nuclear Power Plants, Using Root Cause Analysis, *Applied Ergonomics*, Vol. 20, No. 2, 1989, pp. 115–121.

3. Speaker, D. M., Voska, K. J., Luckas, W. J., *Identification and Analysis of Human Errors Underlying Electric/Electronic Component Related Events*, Report No. NUREG/CR-2987, Nuclear Power Plant Operations, U.S. Nuclear Regulatory Commission, Washington, D.C., 1983.

4. Hobbs, A., Williamson, A., *Human Factors in Airline Maintenance*, Proceedings of the Conference on Applied Psychology, 1995, pp. 384–393.

5. Shepherd, W. T., *The FAA Human Factors Program in Aviation Maintenance and Inspection*, Proceedings of the 5th Federal Aviation Administration (FAA) Meeting on Human Factors Issues in Aircraft Maintenance and Inspection, June 1991, pp. 1–5.

6. Shepherd, W. T., Johnson, W. B., Drury, C. G., Berninger, D., *Human Factors in Aviation Maintenance Phase One: Progress Report No. AM-91/16*, Office of Aviation Medicine, Federal Aviation Administration (FAA), Washington, D.C., November, 1991.

7. Report No. DOC 9824-AN1450, *Human Factors Guidelines for Aircraft Maintenance Manual*, International Civil Aviation Organization (ICAO), Montreal, Canada, 2003.

8. Reason, J., Human Factors in Nuclear Power Generation: A Systems Perspective, *Nuclear Europe Worldscan*, Vol. 17, No. 5–6, 1997, pp. 35–36.

9. *An Analysis of 1990 Significant Events*, Report No. INP91-018, Institute of Nuclear Power Operations (INPO), Atlanta, GA, 1991.

10. Hasegawa, T., Kameda, A., *Analysis and Evaluation of Human Error Events in Nuclear Power Plants*, paper presented at the Meeting of the IAEA's CRP on Collection and Classification of Human Reliability Data for Use in Probabilistic Safety Assessments, May 1998. Available from the Institute of Human Factors, Nuclear Power Engineering Corporation, 3-17-1, Toranomon, Minatoku, Tokyo, Japan.

11. The UAW and the Rouge Explosion: A Pat on the Head, *Detroit News*, Detroit, MI, February 6, 1999, pp. 6.

12. White, J., *New Revelations Expose Company-Union Complexity in Fatal Blast at U.S. Ford Plant*. Available online at http://www.wsws.org/articles/2000/feb 2000/ford 04.shtml.

13. Daniels, R. W., *The Formula for Improved Plant Maintainability Must Include Human Factors*, Proceedings of the IEEE Conference on Human Factors and Nuclear Safety, 1985, pp. 242–244.

14. Maintenance Error a Factor in Blackouts, *Miami Herald*, Miami, FL, December 29, 1989, pp. 4.

15. Pyy, P., An Analysis of Maintenance Failures at a Nuclear Power Plant, *Reliability Engineering and Systems Safety*, Vol. 72, 2001, pp. 293–302.

16. Pyy, P., Laakso, K., Reiman, L., *A Study of Human Errors Related to NPP Maintenance Activities*, Proceedings of the IEEE 6th Annual Human Factors Meeting, 1997, pp. 12.23–12.28.

17. Dhillon, B. S., *Human Reliability, Error, and Human Factors in Engineering Maintenance*, CRC Press, Boca Raton, FL, 2009.

18. Seminara, J. L., Parsons, S. O., *Human Factors Review of Power Plant Maintainability*, Report No. NP-1567 (Research Project 1136), Electric Power Research Institute, Palo Alto, CA, 1981.

19. Seminara, J. L., Parsons, S. O., Human Factors Engineering and Power Plant Maintenance, *Maintenance Management International*, Vol. 6, 1985, pp. 33–71.
20. Isoda, H., Yasutake, J. Y., *Human Factors Interventions to Reduce Human Errors and Improve Productivity in Maintenance Tasks*, Proceedings of the International Conference on Design and Safety of Advanced Nuclear Power Plants, 1992, pp. 34.4-1–34.4-6.
21. Knee, H. E., *The Maintenance Personnel Performance Simulation (MAPPS) Model: A Human Reliability Analysis Tool*, Proceedings of the International Conference on Nuclear Power Plant Aging, Availability Factor and Reliability Analysis, 1985, pp. 77–80.
22. Marx, D. A., *Learning from Our Mistakes: A Review of Maintenance Error Investigation and Analysis Systems* (with recommendations to the FAA), Federal Aviation Administration (FAA), Washington, D.C., January 1998.
23. Russell, P. D., Management Strategies for Accident Prevention, *Air Asia*, Vol. 6, 1994, pp. 31–41.
24. Report No. 2-97, *Human Factors in Airline Maintenance: A Study of Incident Reports*, Bureau of Air Safety Investigation (BASI), Department of Transport and Regional Development, Canberra, Australia, 1997.
25. Report No. CAP 718, *Human Factors in Aircraft Maintenance and Inspection*, Prepared by the Safety Regulation Group, Civil Aviation Authority, London, 2002. Available from the Stationary Office, Norwich, U.K.
26. Graeber, R. C., Max, D. A., *Reducing Human Error in Aircraft Maintenance Operations*, Proceedings of the 46th Annual International Safety Seminar, 1993, pp. 147–160.
27. Latorella, K. A., Prabhu, P. V., A Review of Human Error in Aviation Maintenance and Inspection, *International Journal of Industrial Ergonomics*, Vol. 26, 2000, pp. 133–161.
28. Christensen, J. M., Howard, J. M., Field Experience in Maintenance, in *Human Detection and Diagnosis of System Failures*, eds. J. Rasmussen and W. B. Rouse, Plenum Press, New York, 1981, pp. 111–133.
29. Report No. 93-1, *Investigation of Human Factors in Accidents and Incidents*, International Civil Aviation Organization, Montreal, Canada, 1993.
30. Dhillon, B. S., *Human Reliability: With Human Factors*, Pergamon Press, New York, 1986.
31. Maintenance Error Decision Aid (MEDA), Developed by Boeing Commercial Airline Group, Seattle, WA, 1994.
32. Allen, J. P., Rankin, W. L., *A Summary of the Use and Impact of the Maintenance Error Decision Aid (MEDA) on the Commercial Aviation Industry*, Proceedings of the 48th Annual International Air Safety Seminar, 1995, pp. 359–369.
33. Rankin, W., MEDA Investigation Process, *Aero Quarterly*, Vol. 2, No. 1, 2007, pp. 15–22.
34. Rankin, W. L., Allen, J. P., Sargent, R. A., *Maintenance Error Decision Aid: Progress Report*, Proceedings of the 11th FAA/AAM Meeting on Human Factors in Aviation Maintenance and Inspection, 1997, pp. 19–24.
35. Hibit, R., Marx, D. A., *Reducing Human Error in Aircraft Maintenance Operations with the Maintenance Error Decision Aid (MEDA)*, Proceedings of the Human Factors and Ergonomics Society 38th Annual Meeting, 1994, pp. 111–114.
36. Dhillon, B. S., *Human Reliability and Error in Transportation Systems*, Springer Verlag, London, 2007.

Further Reading: Literature on Safety and Human Error in Engineering Systems

Introduction

Over the years, a large number of publications on safety and human error in engineering systems have appeared in the form of journal articles, conference proceedings articles, technical reports, and so on. This appendix presents an extensive list of selective publications related, directly or indirectly, to safety and human error in engineering systems.

The period covered by the listing is from 1926 to 2009. The main objective of this listing is to provide readers with sources for obtaining additional information on safety and human error in engineering systems.

Publications

1. Abkowitz, M., Availability and Quality of Data for Assessing Heavy Truck Safety, *Transportation Quarterly*, Vol. 44, No. 2, 1990, pp. 203–226.
2. Acobsen, T., A Potential of Reducing the Risk of Ship Casualties by 50%, *Marine and Maritime*, Vol. 3, 2003, pp. 171–181.
3. Adams, S. K., Sabri, Z. A., Husseiny, A. A., *Maintenance and Testing Errors in Nuclear Power Plants: A Preliminary Assessment*, Proceedings of the Human Factors 24th Annual Meeting, 1980, 280–284.
4. AFR-122-9, *Nuclear Surety Design Certification for Nuclear Weapon System Software and Firmware*, Department of the Air Force, Washington, D.C., August 1987.
5. Ahlstrom, U., Work Domain Analysis for Air Controller Weather Displays, *Journal of Safety Research*, Vol. 36, 2005, pp. 159–169.
6. Allen, J. O. Rankin, W. L., Use of the Maintenance Error Decision Aid (MEDA) to *Enhance Safety and Reliability and Reduce Costs in the Commercial Aviation Industry*, Proceedings of the Tenth Federal Aviation Administration Meeting on Human Factors Issues in Aircraft Maintenance and Inspection: Maintenance Performance Enhancement and Technician Resource Management, 1996, pp. 79–87.
7. Allen, J. P., Marx, D. M., *Maintenance Error Decision Aid Project*, Proceedings of the Eighth Federal Aviation Administration Meeting on Human Factors Issues in Aircraft Maintenance and Inspection: Trends and Advances in Aviation Maintenance Operations, 1994, pp. 101–116.
8. Allen, J. P., Rankin, W. L., *Summary of the Use and Impact of the Maintenance Error Decision Aid (MEDA) on the Commercial Aviation Industry*, Proceedings of the International Air Safety Seminar, 1995, pp. 359–369.
9. Amalberti, R., Wioland, L., *Human Error in Aviation*, Proceedings of the International Aviation Safety Conference on Aviation Safety: Human Factors, System Engineering, Flight Operations, Economics, and Strategies Management. 1997, pp. 91–108.

10. Ambs, J. L., Setren, R. S., *Safety Evaluation of Disposable Diesel Exhaust Filters for Permissible Mining Equipment*, Proceedings of the Seventh US Mine Ventilation Symposium, 1995, pp. 105–110.

11. American National Standard for Industrial Robots and Robot Systems-Safety Requirements, ANSI/RIA R 15.06 – 1986, American National Standards Institute (ANSI), New York, 1986.

12. Amrozowlcz, M. D., Brown, A., Golay, M., *Probabilistic Analysis of Tanker Groundings*, Proceedings of the International Offshore and Polar Engineering Conference, Vol. 4, 1997, pp. 313–320.

13. *An Interpretation of the Technical Guidance on Safety Standards in the Use, etc., of Industrial Robots*, Japanese Industrial Safety and Health Association, Tokyo, 1985.

14. Anderson, D. E., Malone, T. B., Baker, C. C., Recapitalizing the Navy through Optimized Manning and Improved Reliability, *Naval Engineers Journal*, Vol. 110, No. 6, 1998, pp. 61–72.

15. Anderson, D. E., Oberman, F. R., Malone, T. B., Baker, C. C., Influence of Human Engineering on Manning Levels and Human Performance on Ships, *Naval Engineers Journal*, Vol. 109, No. 6, 1997, pp. 67–76.

16. Anderson, F. A., Medical Device Risk Assessment, in *The Medical Device Industry*, eds. N. F. Estrin, Marcel Dekker, New York, 1990, pp. 487–493.

17. Anon., Accidents: U.K. Investigators Say Fatal Rail Collision Not Fault of Railroad, *Engineering News Record (ENR)*, Vol. 246, No. 10, 2001, p. 15.

18. Anon., Cargo Collection Points Part of Truck Safety Plan, *AFE Facilities Engineering Journal*, Vol. 29, No. 4, 2002, pp. 11–13.

19. Anon., Casting Concrete Walls to Improve Road and Rail Safety, *Quality Concrete*, Vol. 8, No. 5-6, 2002, p. 31.

20. Anon., (Editorial) Needed: Greater Lift Truck Safety, *Modern Materials Handling*, Vol. 51, No. 2, 1996, pp. 50–53.

21. Anon., Ensuring Railroad Tank Car Safety, *TR News*, No. 176, 1995, pp. 30–31.

22. Anon., Exclusive: Railtrack to Develop Track Fault Sensors for Trains, *Engineer*, Vol. 291, No. 7598, 2002, p. 18.

23. Anon., Great Britain Promotes Rail Safety Research, *TR News*, No. 226, 2003, p. 51.

24. Anon., Minimizing Train Delays Can Improve Mobility, Safety at Rail Grade Crossings, *Texas Transportation Researcher*, Vol. 43, No. 3, 2007, p. 7.

25. Anon., Rail Inspector Slams Industry Safety, *Professional Engineering*, Vol. 12, No. 2, 1999, p. 4.

26. Anon., Rail Rudder Propeller Variant Offers More Safety in Coastal Shipping, *HSB International*, Vol. 41, No. 11, 1993, pp. 40–41.

27. Anon., Railroads Using Science to Advance Safety, *Railway Track and Structures*, Vol. 94, No. 9, 1998, p. 27.

28. Anon., Railway Firms on Track with Safety Systems, *Professional Engineering*, Vol. 14, No. 18, 2001, p. 8.

29. Anon., Research: Rail Polishing Increases Track Life and Train Safety, *Engineering News Record*, Vol. 243, No. 1, 1999, p. 12.

30. Anon., Research to Improve Road/Rail Safety, *Professional Engineering*, Vol. 14, No. 10, 2001, pp. 1–4.

31. Anon., Safety Assessment for Ships Manoeuvring in Ports, *Dock & Harbour Authority*, Vol. 79, No. 889–892, 1998, pp. 26–29.

32. Anon., Safety in Ship Building, *Motor Ship*, Vol. 78, No. 922, 1997, pp. 17–18.

33. Anon., Safety Pole Keeps Trucks at the Dock, *Material Handling Engineering*, Vol. 53, No. 6, 1998, p. 286.

34. Anon., Safety: Heart of the Matter-Bad Cargo Stowage Can Affect the Safety of the Ship, *Cargo Systems*, Vol. 22, No. 9, 1995, pp. 74–76.

35. Anon., Sharing Best Practice: Synergy in Rail and Aviation Safety, *Aircraft Engineering and Aerospace Technology*, Vol. 73, No. 1, 2001, p. 15.

36. Anon., Small Power Tool Makers Strive for Safety to Meet Railroad Demands, *Railway Track and Structures*, Vol. 98, No. 8, 2002, pp. 31–33.

37. Anon., Truck Transportation Safety, *Papermaker*, Vol. 59, No. 3, 1996, p. 3.

38. Anon., Truckers Division: MTU to Study Log Truck Safety, *Timber Producer*, No. 2, 2002, p. 6.

39. Anon., Turning the Roundhouse into a Roundtable: The National Highway-Rail Grade Crossing Safety Training Conference, *Texas Transportation Researcher*, Vol. 44, No. 2, 2008, p. 13.

40. ANSI/AAMI HE-48, *Human Factors Engineering Guidelines and Preferred Practices for the Design of Medical Devices*, Association for the Advancement of Medical Instrumentation (AAMI), Arlington, VA, 1993.

41. Archer, R. D., Lewis, G. W., Lockett, J., *Human Performance Modeling of Reduced Manning Concepts for Navy Ships*, Proceedings of the Human Factors and Ergonomics Society Annual Meeting, Vol. 2, 1996, pp. 987–991.

42. Arearese, J. S., FDA's Role in Medical Device User Education, in *The Medical Device Industry*, Estrin, N. F., Ed., Marcel Dekker, New York, 1990, pp. 129–138.

43. Ayyub, B. M., Beach, J. E., Sarkani, S., Assakkaf, I. A., Risk Analysis and Management for Marine Systems, *Naval Engineers Journal*, Vol. 114, No. 2, 2002, pp. 181–206.

44. Bacchi, M., Cacciabue, C., O' Connor, S., *Reactive and Proactive Methods for Human Factors Studies in Aviation Maintenance*, Proceedings of the Ninth International Symposium on Aviation Psychology, 1997, pp. 991–996.

45. Balsi, M., Racina, N., *Automatic Recognition of Train Tail Signs using CNNs*, Proceedings of the IEEE International Workshop on Cellular Neural Networks and their Applications, 1994, pp. 225–229.

46. Bandyopadhyay, D., Safety Management Ships, *Journal of the Institution of Engineers* (India), Part MR: Marine Engineering Division, Vol. 84, No. 2, 2003, pp. 45–48.

47. Baranyi, E., Racz, G., Szabo, G., Saghi, B., Traffic and Interlocking Simulation in Railway Operation: Theory and Practical Solutions, *Periodica Polytechnica Transportation Engineering*, Vol. 33, No. 1–2, 2005, pp. 177–185.

48. Barczak, T., Gearhart, D., *Performance and Safety Considerations of Hydraulic Support Systems*, Proceedings of the 17th International Conference on Ground Control in Mining, 1998, pp. 176–186.

49. Barkand, T. D, *Emergency Braking Systems For Mine Elevators*, Proceedings of the Conference on New Technology in Mine Health and Safety, 1992, pp. 325–336.

50. Barnes, H. J., Levine, J. D., Wogalter, M. S., *Evaluating the Clarity of Highway Entrance-Ramp Directional Signs*, Proceedings of the XIVth Triennial Congress of the International Ergonomics Association and 44th Annual Meeting of the Human Factors and Ergonomics Association, "Ergonomics for the New Millennium," 2000, pp. 794–797.

51. Bassen, H., Silberberg, J., Houston, F., Knight, W., Christman, C., Greberman, M., *Computerized Medical Devices: Usage, Trends, Problems, and Safety Technology*, Proceedings of the 7th Annual Conference of the IEEE/Engineering in Medicine and Biology Society, 1985, pp. 180–185.

52. Bassen, H., Silberberg, J., Houston, F., Knight, W., Computerized Medical Devices, Trends, Problems, and Safety, *IEEE Aerospace and Electronic Systems (AES) Magazine*, September 1986, pp. 20–24.

53. Bayley, J. M., Uber, C. B., *Comprehensive Program to Improve Road Safety at Railway Level Crossings*, Proceedings of the 15th Australian Road Research Board Conference, 1990, pp. 217–234.

54. Beder, D. M., Romanovsky, A., Randell, B., Snow, C. R., Stroud, R. J., An Application of Fault Tolerance Patterns and Coordinated Atomic Actions to a Problem in Railway Scheduling, *Operating Systems Review (ACM)*, Vol. 34, No. 4, 2000, pp. 21–31.

55. Bennett, C. T., Schwirzke, M., Harm, C., *Analysis of General Aviation Accidents during Operations under Instrument Flight Rules*, Proceedings of the Human Factors and Ergonomics Society Annual Meeting, 1990, pp. 1057–1061.

56. Bercha, F. G., Brooks, C. J., Leafloor, F., *Human Performance in Arctic Offshore Escape, Evacuation, and Rescue*, Proceedings of the International Offshore and Polar Engineering Conference, 2003, pp. 2755–2763.

57. Beus, M. J., Iverson, S., Safer Mine Hoisting with Conveyance Position and Load Monitoring, *American Journal of Industrial Medicine*, Vol. 36, 1999, pp. 119–121.

58. Blache, K. M., Industrial Practices for Robotic Safety, in *Safety, Reliability, and Human Factors*, ed. J. H. Graham, Van Nostrand Reinhold, New York, 1991, pp. 34–65.

59. Blackmon, R. B., Gramopadhye, A. K., *Using the Aircraft Inspector's Training System to Improve the Quality of Aircraft Inspection*, Proceedings of the 5th Industrial Engineering Research Conference, 1996, pp. 447–452.

60. Blekinsop, G., Only Human, *Quality World*, Vol. 29, No. 12, 2003, pp. 24–29.

61. Bobick, T. G., *Increasing Safety in Underground Coal Mining through Improved Materials-Handling Systems*, Proceedings of the Society of Mining Engineers (SME) Annual Meeting, 1988, pp. 1–9.

62. Bob-Manuel, K. D. H., Probabilistic Prediction of Capsize Applied to Small High-Speed Craft, *Ocean Engineering*, Vol. 29, 2002, pp. 1841–1851.

63. Bogner M. S., Medical Device and Human Error, in *Human Performance in Automated Systems: Current Research and Trends*, eds. Mouloua, M., and Parasuraman, P., Lawrence Erlbaum Associates Publishers, Hillsdale, NJ, 1994, pp. 64–67.

64. Bogner, M. S., Designing Medical Devices to Reduce the Likelihood of Error, *Biomedical Instrumentation & Technology*, Vol. 33, No. 2, 1999, pp. 108–113.

65. Boniface, D. E., Bea, R. G., Assessing the Risks of and Countermeasures for Human and Organizational Error, *Transactions of the Society of Naval Architects and Marine Engineers*, Vol. 104, 1996, pp. 157–177.

66. Bonney, M. C., Yong, Y. F., Eds., *Robot Safety*, Springer-Verlag, New York, 1985.

67. Borse, E., Design Basis Accidents and Accident Analysis with Particular Reference to Offshore Platforms, *Journal of Occupational Accidents*, Vol. 2, No. 3, 1979, pp. 227–243.

68. Bos, T., Hoekstra, R., *Reduction of Error Potential in Aircraft Maintenance*, available from the Human Factors Department, National Aerospace Laboratory NLR, 1006 BM Amsterdam, The Netherlands.

69. Bourne, A., Managing Human Factors in London Underground, *IEE Colloquium* (Digest), No. 49, 2000, pp. 5/1–5/3.

70. Bousvaros, G. A., Don, C., Hopps, J. A., An Electrical Hazard of Selective Angiocardiography, *Canadian Medical Association Journal*, Vol. 87, 1962, pp. 286–288.

71. Bradley, E. A., Case Studies in Disaster—a Coded Approach, *International Journal of Pressure Vessels and Piping*, Vol. 61, No. 2–3, 1995, pp. 177–197.

72. Brooker, P., Airborne Separation Assurance Systems: Towards a Work Programme to Prove Safety, *Safety Science*, Vol. 42, No. 8, 2004, pp. 723–754.

73. Brown, A., Haugene, B., *Assessing the Impact of Management and Organizational Factors on the Risk of Tanker Grounding*, Proceedings of the International Offshore and Polar Engineering Conference, Vol. 4, 1998, pp. 469–477.

74. Brown, I. D., Drivers' Margins of Safety Considered as a Focus for Research on Error, *Ergonomics*, Vol. 33, No. 10–11, 1990, pp. 1307–1314.

75. Brown, I. D., Prospects for Technological Countermeasures against Driver Fatigue, *Accident Analysis and Prevention*, Vol. 29, No. 4, 1997, pp. 525–531.

76. Brown, S. L., Use of Risk Assessment Procedures for Evaluating Risks of Ethylene Oxide Residues in Medical Devices, in *The Medical Device Industry*, ed. N. F. Estrin, Marcel Dekker, New York, 1990, pp. 469–485.

77. Bruley, M. E., *Ergonomics and Error—Who Is Responsible?* Proceedings of the First Symposium on Human Factors in Medical Devices, 1989, pp. 6–10.

78. Bruner, J. M. R., Hazards of Electrical Apparatus, *Anesthesiology*, Vol. 28, No. 2, 1967, pp. 396–425.

79. Bu, F., Chan, C., *Pedestrian Detection in Transit Bus Application: Sensing Technologies and Safety Solutions*, Proceedings of the IEEE Intelligent Vehicles Symposium, 2005, pp. 100–105.

80. Buck, L., Error in the Perception of Railway Signals, *Ergonomics*, Vol. 6, 1968, pp. 181–192.

81. Burchell, H. B., Electrocution Hazards in the Hospital or Laboratory, *Circulation*, 1963, pp. 1015–1017.

82. Burlington, D. B., Human Factors and the FDA's Goal: Improved Medical Device Design, *Biomedical Instrumentation & Technology*, Vol. 30, No. 2, 1996, pp. 107–109.

83. Butani, S. J., *Hazard Analysis of Mining Equipment by Mine Type and Geographical Region*, Proceedings of the Symposium on Engineering Health and Safety, 1986, pp. 158–173.

84. Butsuen, T., Yoshioka, T., Okuda, K., *Introduction of the Mazda Advanced Safety Vehicle*, Proceedings of the IEEE Intelligent Vehicles Symposium, 1996, pp. 242–249.

85. Cacciabue, P. C., Human Error Risk Management Methodology for Safety Audit of a Large Railway Organisation, *Applied Ergonomics*, Vol. 36, No. 6, 2005, pp. 709–718.

86. Cafiso, S., Condorelli, A., Cutrona, G., Mussumeci, G., A Seismic Network Reliability Evaluation on a GIS Environment—A Case Study on Catania Province, *Management Information Systems*, Vol. 9, 2004, pp. 131–140.

87. Callantine, T. J., *Agents for Analysis and Design of Complex Systems*, Proceedings of the IEEE International Conference on Systems, Man and Cybernetics, 2001, pp. 567–573.

88. Callantine, T. J., *Air Traffic Controller Agents*, Proceedings of the International Conference on Autonomous Agents, Vol. 2, 2003, pp. 952–953.

89. Camishion, R. C., Electrical Hazards in the Research Laboratory, *Journal of Surgical Research*, Vol. 6, 1966, pp. 221–227.

90. Carey, M. S., Delivering Effective Human Systems, *IEE Colloquium* (Digest), No. 49, April 2000, pp. 6/–6/5.

91. Cartmale, K., Forbes, S. A., Human Error Analysis of a Safety Related Air Traffic Control Engineering Procedure, *IEE Conference Publication*, No. 463, 1999, pp. 346–351.

92. Castaldo, R., Evers, C., Smith, A., *Improved Location/Identification of Aircraft/ Ground Vehicles on Airport Movement Areas: Results of FAA Trials*, Proceedings of the Institute of Navigation National Technical Meeting, 1996, pp. 555–562.

93. Cawley, J. C., *Electrical Accidents in the Mining Industry, 1990–1999*, Proceedings of the Thirty-Sixth IAS Annual Meeting, 2001, Vol. 2, pp. 1361–1368.

94. Cawley, J. C., *Probability of Spark Ignition in Intrinsically Safe Circuits*, Report No. 9183, Report of Investigations, U.S. Bureau of Mines, Washington, D.C., 1988, pp. 1–15.

95. Cawley, J. C., Electrical Accidents in the Mining Industry, 1990–1999, *IEEE Transactions on Industry Applications*, Vol. 39, No. 6, 2003, pp. 1570–1577.

96. Cha, S. S., *Management Aspect of Software Safety*, Proceedings of the Eighth Annual Conference on Computer Assurance, 1993, pp. 35–40.

97. Chan, K., Turner, D., The Application of Selective Door Opening within a Railway System, *Advances in Transport*, Vol. 15, 2004, pp. 155–164.

98. Chang, C. S., Lau, C. M., Design of Modern Control Centres for the 21st Century—Human Factors and Technologies, *IEE Conference Publication*, No. 463, 1999, pp. 131–136.

99. Chang, C. S., Livingston, A. D., Chang, D., Achieving a Uniform and Consistent Graphical User Interface for a Major Railway System, *Advances in Transport*, Vol. 15, 2004, pp. 187–197.

100. Chen, S., Gramopadhye, A., Melloy, B., *The Effects of Individual Differences and Training on Paced and Unpaced Aircraft Visual Inspection Performance*, Proceedings of the XIVth Triennial Congress of the International Ergonomics Association and 44th Annual Meeting of the Human Factors and Ergonomics Society, 2000, pp. 491–494.

101. Chevlin, D. H., Jorgens, J., Medical Device Software Requirements: Definition and Specification, *Medical Instrumentation*, Vol. 30, No. 2, March/April 1996.

102. Clark, D. R., Lehto, M. R., Reliability, Maintenance, and Safety of Robots, in *Handbook of Industrial Robotics*, ed. S. Y. Nof, John Wiley & Sons, New York, 1999, pp. 717–753.

103. Clark, R. D., Bonney, S., *Integration of Human Factors and System Safety in the Collapse Analysis of Roll-Over Safety Rings in Buses and Coaches*, International Conference on Traffic and Transportation Studies, 1998, pp. 820–829.

104. Collins, E. W., *Safety Evaluation of Coal Mine Power Systems*, Proceedings of the Annual Reliability and Maintainability Symposium, 1987, pp. 51–56.

105. Congress, N., Automated Highway System: An Idea Whose Time Has Come, *Public Roads*, Vol. 58, No. 1, 1994, pp. 1–9.

106. Cooper, J. B., Newbower, R. S., Kitz, R. J., An Analysis of Major Errors and Equipment Failures in Anaesthesia Management: Consideration for Prevention and Detection, *Anaesthesiology*, Vol. 60, No. 1, 1984, pp. 34–42.

107. Cothen, G. C., Schulte, C. F., Horn, J. D., Tyrell, D. C., Consensus Rulemaking at the Federal Railroad Administration: All Aboard for Railway Safety Measures, *TR News*, No. 236, 2005, pp. 8–14.

108. Cox, T., Houndmont, J., Griffiths, A., Rail Passenger Crowding, Stress, Health, and Safety in Britain, *Transportation Research Part A: Policy and Practice*, Vol. 40, No. 3, 2006, pp. 244–358.

109. Cunningham, B. G., *Maintenance Human Factors at Northwest Airlines*, Proceedings of the 10th Federal Aviation Administration, Meeting on Human Factors Issues in Aircraft Maintenance and Inspection: Maintenance Performance Enhancement and Technician Resource Management, 1996, pp. 43–53.

110. Danahar, J. W., *Maintenance and Inspection Issues in Aircraft Accidents/Incidents*, Proceedings of the Meeting on Human Factors Issues in Aircraft Maintenance and Inspection, 1989, pp. A9–A11.

111. Daniel, J. H., *Reducing Mine Accidents by Design*, Proceedings of the SME Annual Meeting, 1991, pp. 1–11.

112. Davies, R. K. L., Monitoring Hoist Safety Equipment, *Coal Age*, Vol. 21, No. 1, 1986, pp. 66–68.

113. Dawes, S. M., Integrated Framework to Analyze Coordination and Communication among Aircrew, Air Traffic Control, and Maintenance Personnel, *Transportation Research Record*, No. 1480, 1995, pp. 9–16.

114. Day, L. M., Farm Work Related Fatalities among Adults in Victoria, Australia, the Human Cost of Agriculture, *Accident Analysis and Prevention*, Vol. 31, No. 1–2, 1998, pp. 153–159.

115. De Groot, H., *Flight Safety: A Human Factors Task*, Proceedings of the International Air Safety Seminar, 1990, pp. 102–106.

116. De Rosa, M. I., Equipment Fires Cause Injuries, *Coal Age*, Vol. 109, No. 10, 2004, pp. 28–31.

117. Deierlein, B., Spec'ing Trucks for Safety, *Waste Age*, Vol. 32, No. 3, 2001, pp. 212–216.

118. Denzler, H. E., How Safety Is Designed into Offshore Platforms, *World Oil*, Vol. 152, No. 7, 1961, pp. 131–133, 136.

119. Dhananjay, K., Bengt, K., Uday, K., Reliability Analysis of Power Transmission Cables of Electric Mine Loaders Using the Proportional Hazards Model, *Reliability Engineering & System Safety*, Vol. 37, No. 3, 1992, pp. 217–222.

120. DHHS (NIOSH) Publication No. 85-103, *Preventing the Injury of Workers by Robots*, National Institute for Occupational Safety and Health (NIOSH), Morgantown, WV, 1984.

121. Dhillon, B. S., Fashandi, A. R. M., Safety and Reliability Assessment Techniques in Robotics, *Robotica*, Vol. 15, 1997, pp. 701–708.

122. Dhillon, B. S., Fashandi, A R. M., Stochastic Analysis of a Robot Machine with Duplicate Safety Units, *Journal of Quality in Maintenance Engineering*, Vol. 5, No. 2, 1999, pp. 114–127.

123. Dhillon, B. S., *Human Error in Medical Systems*, Proceedings of the 6th ISSAT International Conference on Reliability and Quality in Design, 2000, pp. 138–143.

124. Dhillon, B. S., *Robot Reliability and Safety*, Springer-Verlag, New York, 1991.

125. Dhillon, B. S., Tools for Improving Medical Equipment Reliability and Safety, *Physics in Medicine and Biology*, Vol. 39a, 1994, p. 941.

126. Dhillon, B. S., Yang, N., Formulas for Analyzing a Redundant Robot Configuration with a Built-in Safety System, *Microelectronics and Reliability*, Vol. 37, No. 4, 1997, pp. 557–563.

127. Di Benedetto, M. D., Di Gennaro, S., D'Innocenzo, A., *Critical Observability and Hybrid Observers for Error Detection in Air Traffic Management*, Proceedings of the IEEE International Symposium on Intelligent Controls, Vol. 2, 2005, pp. 1303–1308.

128. Diehl, A., *Effectiveness of Aeronautical Decision Making Training*, Proceedings of the Human Factors Society Annual Meeting, 1990, pp. 1367–1371.

129. Dieudonne, J., Joseph, M., Cardosi, K., *Is the Proposed Design of the Aeronautical Data Link System Likely to Reduce the Miscommunications Error Rate and Controller/ Flight Crew Input Errors?* Proceedings of the AIAA/IEEE 13 Digital Avionics Systems Conference, Vol. 2, 2000, pp. 5.E.3.1–5.E.3.9.

130. Doll, R., Maintenance and Inspection Issues in Air Carrier Operations, in *Human Factors Issues in Aircraft Maintenance and Inspection*, Report No. DOT/FAA/ AAM-89/9, Office of Aviation Medicine, Federal Aviation Administration, Washington, D.C., 1989, pp. A33–36.

131. Donelson, A. C., Ramachandran, K., Zhao, K., Kalinowski, A., Rates of Occupant Deaths in Vehicle Rollover: Importance of Fatality-Risk Factors, *Transportation Research Record*, No. 1665, 1999, pp. 109–117.

132. Dorn, M. D., Effects of Maintenance Human Factors in Maintenance—Related Aircraft Accidents, *Transportation Research Record*, No. 1517, 1996, pp. 17–28.

133. Douglass, D. P., Safety Devices on Electric Hoists Used in Ontario Mines, *Canadian Mining Journal*, Vol. 61, No. 4, 1940, pp. 229–234.

134. Drury, C. G., *Integrating Training into Human Factors Implementation*, Proceedings of the Human Factors and Ergonomics Society Annual Meeting, Vol. 2, 1996, pp. 1082–1086.

135. Drury, C. G., *Errors in Aviation Maintenance: Taxonomy and Control*, Proceedings of the 35th Annual Meeting of the Human Factors Society, 1991, pp. 42–46.

136. Drury, C. G., Murthy, M. R., Wenner, C. L., *A Proactive Error Reduction System*, Proceedings of the 11th Federal Aviation Administration Meeting on Human Factors Issues in Aircraft Maintenance and Inspection: Human Error in Aviation Maintenance, 1997, pp. 91–103.

137. Drury, C. G., Prabhu, P., Gramopadhye, A., *Task Analysis of Aircraft Inspection Activities. Methods and Findings*, Proceedings of the Human Factors Society Annual Meeting, 1990, pp. 1181–1185.

138. Drury, C. G., Rangel, J., Reducing Automation—Related Errors in Maintenance and Inspection, in *Human Factors in Aviation Maintenance*—Phase VI: Progress Report, Vol. II, Office of Aviation Medicine, Federal Aviation Administration, Washington, D.C., 1996, pp. 281–306.

139. Drury, C. G., Sarac, A., *A Design Aid for Improved Documentation in Aircraft Maintenance: A Precursor to Training*, Proceedings of the Human Factors and Ergonomics Society Annual Meeting, Vol. 2, 1997, pp. 1158–1162.

140. Drury, C. G., Shepherd, W. T., Johnson, W. B., *Error Reduction in Aviation Maintenance*, Proceedings of the 13th Triennial Congress of the International Ergonomics Association, 1997, pp. 31–33.

141. Drury, C. G., Spencer, F. W., *Measuring Human Reliability in Aircraft Inspection*, Proceedings of the 13th Triennial Congress of the International Ergonomics Association, 1997, pp, 34–36.

142. Drury, C. G., Wenner, C. L., Murthy, M., *A Proactive Error Reduction System*, Proceedings of the 11th FAA/AAM Meeting on Human Factors in Aviation Maintenance and Inspection, 1997, pp. 93–106.

143. Drury, C. G., Wenner, C. L., Murthy, M., A Proactive Error Reporting System, in *Human Factors in Aviation Maintenance*—Phase VII: Progress Report, Office of Aviation Medicine, Federal Aviation Administration, Washington, D.C., 1997, pp. 173–184.

144. Duffey, R. B., Saull, J. W., Errors in Technological Systems, *Human Factors and Ergonomics in Manufacturing*, Vol. 13, No. 4, 2003, pp. 279–291.

145. DuPont, G., *The Dirty Dozen Errors in Maintenance*, Proceedings of the 11th FAA/ AAM Meeting on Human Factors in Aviation Maintenance and Inspection, 1997, pp. 49–52.

146. Edkins, G. D., Pollock, C. M., Influence of Sustained Attention on Railway Accidents, *Accident Analysis and Prevention*, Vol. 29, No. 4, 1997, pp. 533–539.

147. Egorov, G. V., Kozlyakov, V. V., *Investigation of Coastal and Short Sea Ship's Risk and Hull's Reliability*, Proceedings of the International Conference on Offshore Mechanics and Arctic Engineering, Vol. 2, 2001, pp. 49–54.

148. Eisenberg, P., *Computer/User Interface Design Specification for Medical Devices*, Proceedings of the 6th Annual IEEE Symposium on Computer-Based Medical Systems, 1993, pp. 177–182.

149. El Koursi, E., Flahaut, G., Zaalberg, H., Hessami, A., *Safety Assessment of European Rail Rules for Operating ERTMS*, Proceedings of the International Conference on Automated People Movers, 2001, pp. 811–815.

150. Elahi, B. J., *Safety and Hazard Analysis for Software-Controlled Medical Devices*, Proceedings of the 6th Annual IEEE Symposium on Computer-Based Medical Systems, 1993, pp. 10–15.

151. El-Badan, A., Leheta, H. W., Abdel-Nasser, Y., Moussa, M. A., Safety Assessment for Ship Hull Girders Taking Account of Corrosion Effects, *Alexandria Engineering Journal*, Vol. 41, No. 1, 2002, pp. 71–81.

152. Elliott, L., Mojdehbakhsh, R., *A Process for Developing Safe Software*, Proceedings of the 7th Symposium on Computer-Based Medical Systems, 1994, pp. 241–246.

153. Embrey, D. E., Incorporating Management and Organisational Factors into Probabilistic Safety Assessment, *Reliability Engineering & System Safety*, Vol. 38, No. 1–2, 1992, pp. 199–208.

154. Endsley, M. R., Robertson, M. M., Situation Awareness in Aircraft Maintenance Teams, *International Journal of Industrial Ergonomics*, Vol. 26, 2000, pp. 301–325.

155. Endsley, M. R., Robertson, M. M., *Team Situation Awareness in Aviation Maintenance*, Proceedings of the 10th Federal Aviation Administration Meeting on Human Factors Issues in Aircraft Maintenance and Inspection: Maintenance Performance Enhancement and Technician Resource Management, 1996, pp. 95–101.

156. Endsley, M. R., Rodgers, M. D., *Attention Distribution and Situation Awareness in Air Traffic Control*, Proceedings of the Human Factors and Ergonomics Society Annual Meeting, Vol. 1, 1996, pp. 82–85.

157. Ericson, C. A., *Software and System Safety*, Proceedings of the 5th International System Safety Conference, 1981, pp. III B.1–III B.11.

158. Fahlgren, G., Hagdahl, R., *Complacency*, Proceedings of the International Air Safety Seminar, 1990, pp. 72–76.

159. Fancher, P. S., *Safety Issues Associated with Trucks with Multiple Trailers*, Proceedings of the Winter Annual Meeting of the American Society of Mechanical Engineers, 1991, pp. 11–15.

160. Feng, Z., Xu, Z., Wang, L., Shen, Y., Sun, H., Wang, N., *Driver Error Analysis and Risk Model of Driver-Error of Chinese Railways*, Proceedings of the International Symposium on Safety Science and Technology, 2004, pp. 2113–2117.

161. Fischer, A. L., Camera Sensors Boost Safety for Japanese Railway, *Photonics Spectra*, Vol. 38, No. 9, 2004, pp. 33–34.

162. Fitzpatrik, J., Wright, M., *Qantas Engineering and Maintenance Human Factors: The Human Error and Accident Reduction (HEAR) Programme*, Proccedings of the 11th FAA/AAM Meeting on Human Factors in Aviation Maintenance and Inspection, 1997, p. 25–34.

163. Ford, T., Three Aspects of Aerospace Safety—Human Factors in Airline Maintenance, *Aircraft Engineering and Aerospace Technology*, Vol. 69, No. 3,1997, pp. 262–264.

164. Fortier, S. C., Michael, J. B., *A Risk-Based Approach to Cost-Benefit Analysis of Software Safety Activities*, Proceedings of the Eighth Annual Conference on Computer Assurance, 1993, pp. 53–60.

165. Fox, D., Robotic Safety, *Robotics World*, January/February 1999, pp. 26–29.

166. Freihrr, G., Safety Is Key to Product Quality, *Productivity, Medical Device & Diagnostic Industry Magazine*, Vol. 19, No. 4, 1997, pp. 18–19.

167. Fries, R. C., Pienkowski, P., Jorgens, J., Safe, Effective and Reliable Software Design and Development for Medical Devices, *Medical Instrumentation*, Vol. 30, No. 2, 1996, pp. 75–80.

168. Fujii, Y., Kawabe, H., Iijima, K., Yao, T., *Comparison of Safety Levels of Ship's Hull Girders in Longitudinal Bending Designed by Different Criteria*, Proceedings of the International Offshore and Polar Engineering Conference, 2007, pp. 3692–3698.

169. Fuller, D. A., *Managing Risk in Space Operations: Creating and Maintaining a High Reliability Organization*, Proceedings of the AIAA Space Conference, 2004, pp. 218–223.

170. Fuller, R., Learning To Make Errors. Evidence from a Driving Task Simulation, *Ergonomics*, Vol. 33, No. 10–11, 1990, pp. 1241–1250.

171. Fulton, N. L., Airspace Design: A Conflict Avoidance Model Emphasizing Pilot Communication and Perceptual Capabilities, *Aeronautical Journal*, Vol. 103, No. 1020, 1999, pp. 65–74.

172. Genova, R., Galaverna, M., Sciutto, G., Zavatoni, V., *Techniques for Human Performance Analysis in Railway Applications*, Proceedings of the International Conference on Computer Aided Design, Manufacture and Operation in The Railway and Other Advanced Mass Transit Systems, 1998, pp. 959–968.

173. Gibson, C. S., Aspects of safety affecting mechanical and electrical services in mines, *Canadian Mining Journal*, Vol. 73, No. 5, 1952, pp. 66–71.

174. Gordon, R. P. E., Flin, R. H., Mearns, K., Fleming, M. T., *Assessing the Human Factors Causes of Accidents in the Offshore Oil Industry*, Proceedings of the International Conference on Health, Safety, and Environment in Oil Gas Exploration and Production, 1996, pp. 635–644.

175. Gorzalczynski, S., Limitations of Safety Arrest Mechanisms for Mine Shaft Conveyances, *CIM Bulletin*, Vol. 95, No. 1065, 2002, pp. 67–71.

176. Gowen, L. D., *Specifying and Verifying Safety-Critical Software Systems,* Proceedings of the IEEE 7th Symposium on Computer-Based Medical Systems, 1994, pp. 235–240.

177. Gowen, L. D., Yap, M. Y., *Traditional Software Development's Effects on Safety*, Proceedings of the 6th Annual IEEE Symposium on Computer-Based Medical Systems, 1993, pp. 58–63.

178. Graeber, R. C., Marx, D. A., *Reducing Human Error in Aircraft Maintenance Operations*, Proceedings of the 46th Annual International Air Safety Seminar, and 23rd International Federation of Airworthiness Conference, November 8–11, 1993, pp. 147–157.

179. Graeber, R. C., *Fatigue in Long-Haul Operations. Sources and Solutions*, Proceedings of the International Air Safety Seminar, 1990, pp. 246–257.

180. Graeber, R. C., Moodi, M. M., *Understanding Flight Crew Adherence to Procedures: The Procedural Event Analysis Tool (PEAT)*, Proceedings of the International Air Safety Seminar, 1998, pp. 415–424.

181. Graham, J. H., Overview of Robot Safety, Reliability, and Human Factors Issues, in *Safety, Reliability, and Human Factors in Robotic Systems*, ed. J. H. Graham, Van Nostrand Reinhold, New York, 1991, pp. 1–10.

182. Gramopadhye, A. K., Drury, C. G., Human Factors in Aviation Maintenance: How We Get to Where We Are? *International Journal of Industrial Ergonomics*, Vol. 26, No. 2, 2000, pp. 125–131.

183. Gramopadhye, A. K., Drury, C. G., Prabhu, P., Training for Aircraft Visual Inspection, *Human Factors and Ergonomics in Manufacturing*, Vol. 3, 1997, pp. 171–196.

184. Gramopadhye, A. K., Melloy, B., Him, H., Koenig, S., Nickles, G., Kaufman, J., Thaker, J., Bingham, J., Fowler, D., *ASSIST: A Computer-Based Training Program for Aircraft Inspectors*, Proceedings of the Human Factors and Ergonomics Society Annual Meeting, Vol. 2, 1998, pp. 1644–1650.

185. Grant, J. S., Concepts of Fatigue and Vigilance in Relation to Railway Operation, *Ergonomics*, Vol. 14, 1971, pp. 111–118.

186. Grant, L. J., Regulations and Safety in Medical Equipment Design, *Anaethesia*, Vol. 53, 1998, pp. 1–3.

187. Grubb, N. S., *Inspection and Maintenance Issues in Commuter Air Carrier Operations*, Proceedings of the Meeting on Human Factors Issues in Aircraft Maintenance and Inspection, 1989, pp. A37–A41.

188. Gruber, J., Die Mensch-Maschine-Schnittstelle Im Zusammenhang Mit Der Zuverlaessigkeit Des Systems; (Man-Machine Interface and Its Impact on System Reliability), *ZEV-Zeitschrift fuer Eisenbahnwesen und Verkehrstechnik (Journal for Railway and Transport)*, Vol. 124, No. 2–3, 2000, pp. 103–108.

189. Guo, C., Zhang, D., Li, J. Application of FSA to Loading/Discharging Course of Ship, *Dalian Ligong Daxue Xuebao/Journal of Dalian University of Technology*, Vol. 42, No. 5, 2002, pp. 564–569.

190. Haga, S., An Experimental Study of Signal Vigilance Errors in Train Driving, *Ergonomics*, Vol. 27, 1984, pp. 755–765.

191. Haile, J., Clarke, T., Safety Risk and Human Error—the West Coast Route Modernisation, *IEE Colloquium* (Digest), No. 49, 2000, pp. 4/1–4/9.

192. Hal, D., *The Role of Human Factors Training and Error Management in the Aviation Maintenance Safety System*, Proceedings of the Flight Safety Foundation Annual International Air Safety Seminar, 2005, pp. 245–249.

193. Hale, A. R., Stoop, J., Hommels, J., Human Error Models as Predictors of Accident Scenarios for Designers in Road Transport Systems, *Ergonomics*, Vol. 33, No. 10–11, 1990, pp. 1377–1387.

194. Hamilton, S., Top Truck and Bus Safety Issues, *Public Roads*, Vol. 59, No. 1, 1995, pp. 1–5.

195. Hamilton, W. I., Clarke, T., Driver Performance Modelling and Its Practical Application to Railway Safety, *Applied Ergonomics*, Vol. 36, No. 6, 2005, pp. 661–670.

196. Han, L. D., *Simulating ITS Operations Safety with Virtual Reality*, Proceedings of the Transportation Congress, Vol. 1, 1995, pp. 215–226.

197. Hansen, M., Zhang, Y., Safety of Efficiency: Link between Operational Performance and Operational Errors in the National Airspace System, *Transportation Research Record*, No. 1888, 2004, pp. 15–21.

198. Hansen, M. D., *Survey of Available Software-Safety Analysis Techniques,* Proceedings of the Annual Reliability and Maintainability Symposium, 1989, pp. 46–49.

199. Hanson, E. K. S., *Focus of Attention and Pilot Error*, Proceedings of the Eye Tracking Research and Applications Symposium, 2004, pp. 60–61.

200. Harrald, J. R., Mazzuchi, T. A., Spahn, J., Van Dorp, R., Merrick, J., Shrestha, S., Grabowski, M., Using System Simulation to Model the Impact of Human Error in a Maritime System, *Safety Science*, Vol. 30, No. 1–2, 1998, pp. 235–247.

201. Harrison, M. J., Runway Incursions and Airport Surface Traffic Automation, *SAE (Society of Automotive Engineers) Transactions*, Vol. 100, 1991, pp. 2423–2426.

202. Hartman, H. L., Novak, T., Gregg, A. J., Health Hazards of Diesel and Electric Vehicles in an Underground Coal Mine, *Mining Science & Technology*, Vol. 5, No. 2, 1987, pp. 131–151.

203. Hee, D. D., Pickrell, B. D., Bea, R. G., Roberts, K. H., Williamson, R .B., Safety Management Assessment System (SMAS): A Process for Identifying and Evaluating Human and Organization Factors in Marine System Operations with Field Test Results, *Reliability Engineering and System Safety*, Vol. 65, No. 2, 1999, pp. 125–140.

204. Hee, D. D., Pickrell, B. D., Bea, R. G., Roberts, K. H., Williamson, R. B., Safety Management Assessment System (SMAS): A Process for Identifying and Evaluating Human and Organization Factors in Marine System Operations with Field Test Results, *Reliability Engineering and System Safety*, Vol. 65, No. 2, 1999, pp. 125–140.

205. Heinrich, D. J., *Safer Approaches and Landings: A Multivariate Analysis of Critical Factors,* Proceedings of the Corporate Aviation Safety Seminar, 2005, pp. 103–155.

206. Helmreich, R. L., Managing Human Error in Aviation, *Scientific American*, May 1997, pp. 62–64.

207. Hetzler, W. E., Hirsh, G. L., *Machine Operator Crushed by Robotic Platform,* Nebraska Fatality Assessment and Control Evaluation (FACE) Investigation Report No. 99NE017, The Nebraska Department of Labor, Omaha, October 25, 1999.

208. Heyns, F., Van Der Westhuizen, J., A Mining Case Study: The Safe Maintenance of Underground Railway Track, *Civil Engineering*, Vol. 14, No. 5, 2006, pp. 8–10.

209. Heyns, F. J., Construction and Maintenance of Underground Railway Tracks to Safety Standard of SANS: 0339, *Journal of the South African Institute of Mining and Metallurgy*, Vol. 106, No. 12, 2006, pp. 793–798.

210. Hibit, R., Marx, D. A., *Reducing Human Error in Aircraft Maintenance Operations with the Maintenance Error Decision Aid (MEDA),* Proceedings of the Human Factors and Ergonomics Society 38th Annual Meeting, 1994, pp. 111–114.

211. Hidaka, H., Yamagata, T., Suzuki, Y., Structuring a New Maintenance System, *Japanese Railway Engineering*, No. 132–133, 1995, pp. 7–10.

212. Hinchey, M., *Potential for Ship Control*, Proceedings of the International Conference on Offshore Mechanics and Arctic Engineering, Vol. 1, 1993, pp. 245–248.

213. Hobbs, A., Maintenance Mistakes and System Solutions, *Asia Pacific Air Safety*, Vol. 21, 1999, pp. 1–7.

214. Hobbs, A., Robertson, M. M., *Human Factors in Aircraft Maintenance Workshop Report*, Proceedings of the Third Australian Aviation Psychology Symposium, 1995, pp. 468–474.

215. Hobbs, A., Williamson, A., Associations between Errors and Contributing Factors in Aircraft Maintenance, *Human Factors*, Vol. 42, No. 2, 2003, pp. 186–201.

216. Hobbs, A., Williamson, A., *Human Factors in Airline Maintenance*, Proceedings of the Australian Aviation Psychology Symposium, 1995, pp. 384–393.

217. Hobbs, A., Williamson, A., Skills, Rules and Knowledge in Aircraft Maintenance: Errors in Context, *Ergonomics*, Vol. 45, No. 4, 2002, pp. 290–308.

218. Hong, Y., Changchun, L., Min, X., Tong, G., Yao, M., *Human Reliability Analysis on Ship Power System Control Room Design*, Proceedings of the XIVth Triennial Congress of the International Ergonomics Association and 44th Annual Meeting of the Human Factors and Ergonomics Society, 2000, pp. 537–540.

219. Hopkin, V. D., Safety and Human Error in Automated Air Traffic Control, *IEE Conference Publication*, No. 463, 1999, pp. 113–118.

220. Hopps, J. A., *Electrical Hazards in Hospital Instrumentation*, Proceedings of the Annual Symposium on Reliability, 1969, pp. 303–307.

221. Hou, Y. B., Wang, Y., Gao, Y., Du, J. Y., Wang, M., Zhang, H. F., Jiao, S. L., *The Study of Fault Location and Safety Control for the Mine Ventilating Fan*, Proceedings of the International Conference on Machine Learning and Cybernetics, 2006, pp. 3632–3636.

222. Huang, H., Yuan, X., Yao, X., *Fuzzy Fault Tree Analysis of Railway Traffic Safety*, Proceedings of the Conference on Traffic and Transportation Studies, 2000, pp. 107–112.

223. Huang, W. G., Zhang, L., Cause Analysis and Preventives for Human Error Events in Daya Bay NPP, *Dongli Gongcheng/Nuclear Power Engineering*, Vol. 19, No. 1, 1998, pp. 64–67, 76.

224. Hudoklin, A., Rozman, V., Human Errors versus Stress, *Reliability Engineering & System Safety*, Vol. 37, 1992, pp. 231–236.

225. Hudoklin, A., Rozman, V., Reliability of Railway Traffic Personnel, *Reliability Engineering & System Safety*, Vol. 52, 1996, pp. 165–169.

226. Hudoklin, A., Rozman, V., Safety Analysis of the Railway Traffic System, *Reliability Engineering & System Safety*, Vol. 37, 1992, pp. 7–13.

227. Hughes, S., Warner Jones, S., Shaw, K., Experience in the Analysis of Accidents and Incidents Involving the Transport of Radioactive Materials, *Nuclear Engineer*, Vol. 44, No. 4, 2003, pp. 105–109.

228. Hyman, W. A., Errors in the Use of Medical Equipment, in *Human Error in Medicine*, ed. M. S. Bogner, Lawrence Erlbaum Associates Publishers, Hillsdale, NJ, 1994, pp. 327–348.

229. Hyman, W. A., Human Factors in Medical Devices, in *Encyclopedia of Medical Devices*, ed. J. G. Webster, John Wiley & Sons, New York, 1988, pp. 1542–1553.

230. IEC 601-1: *Safety of Medical Electrical Equipment*, Part 1: General Requirements, International Electrotechnical Commission (IEC), Geneva, 1977.

231. IEEE 1228-1994, *Software Safety Plans*, Institute of Electrical and Electronic Engineers (IEEE), New York, May 1994.

232. IEEE-STD-1228, Standard for Software Safety Plans, Institute of Electrical and Electronic Engineers (IEEE), New York, 1994.

233. Ikeda, T., Human Factors Concerning Drivers of High-Speed Passenger Trains, *Rail International*, No. 3, 1995, pp. 19–24.

234. Industrial Robots and Robot System Safety, Chap. 4, in *OSHA Technical Manual*, Occupational Safety and Health Administration (OSHA), Department of Labor, Washington, D.C., 2001.

235. Ingleby, M., Mitchell, I., *Proving Safety of a Railway Signalling System Incorporating Geographic Data*, Proceedings of the IFAC Symposium, 1992, pp. 129–134.

236. Inoue, T., Kusukami, K., Kon-No, S., *Car Driver Behavior in Railway Crossing Accident*, Quarterly Report of RTRI (Railway Technical Research Institute of Japan), Vol. 37, No. 1, 1996, pp. 26–31.

237. Ippolito, L. M., Wallace, D. R., *A Study on Hazard Analysis in High Integrity Software Standards and Guidelines*, Report No. NISTIR 5589, National Institute of Standards and Technology, U.S. Department of Commerce, Washington, D.C., January 1995.

238. Iskander, W. H., Nutter, R. S., Methodology Development for Safety and Reliability Analysis for Electrical Mine Monitoring Systems, *Microelectronics and Reliability*, Vol. 28, No. 4, 1988, pp. 581–597.

239. Isoda, H., Yasutake, J. Y., *Human Factors Interventions to Reduce Human Errors and Improve Productivity in Maintenance Tasks*, Proceedings of the International Conference on Design and Safety of Advanced Nuclear Power Plants, 1992, pp. 34.4/1–6.

240. Itoh, K., Tanaka, H., Seki, M., *Eye-Movement Analysis of Track Monitoring Patterns of Night Train Operators: Effects of Geographic Knowledge and Fatigue,* Proceedings of the XIVth Triennial Congress of the International Ergonomics Association and 44th Annual Meeting of the Human Factors and Ergonomics Society, 2000, pp. 360–363.

241. Ivaturi, S., Gramopadhye, A. K., Kraus, D., Blackmon, R., *Team Training to Improve the Effectiveness of Teams in the Aircraft Maintenance Environment,* Proceedings of the Human Factors and Ergonomics Society 39th Annual Meeting, 1995, pp. 1355–1359.

242. Jacobsen, T., A Potential of Reducing the Risk of Ship Casualties by 50%, *Marine and Maritime*, Vol. 3, 2003, pp. 171–181.

243. Jacobsson, L., Svensson, O., *Psychosocial Work Strain of Maintenance Personnel during Annual Outage and Normal Operation in a Nuclear Power Plant*, Proceedings of the Human Factors Society 35th Annual Meeting, Vol. 2, 1991, pp. 913–917.

244. Janota, A., Using Z Specification for Railway Interlocking Safety, *Periodica Polytechnica Transportation Engineering*, Vol. 28, No. 1–2, 2000, pp. 39–53.

245. Ji, Q., Zhu, Z., Lan, P., Real-Time Nonintrusive Monitoring and Prediction of Driver Fatigue, *IEEE Transactions on Vehicular Technology*, Vol. 53, No. 4, 2004, pp. 1052–1068.

246. Jiang, B. C., Gainer, C. A., A Cause and Effect Analysis of Robot Accidents, *Journal of Occupational Accidents*, Vol. 9, 1987, pp. 27–45.

247. Joel, K., Duncan, S., A Practical Approach to Fire Hazard Analysis for Offshore Structures, *Journal of Hazardous Materials*, Vol. 104, No. 1–3, 2003, pp. 107–122.

248. Johnson, W. B, Rouse, W. B., Analysis and Classification of Human Error in Troubleshooting Live Aircraft Power Plants, *IEEE Transactions on Systems, Man, and Cybernetics,* Vol. 12, No. 3, 1982, pp. 389–393.

249. Johnson, W. B., *National Plan for Aviation Human Factors: Maintenance Research Issues*, Proceedings of the Human Factors Society Annual Meeting, 1991, pp. 28–32.

250. Johnson, W. B., Norton, J. E., *Using Intelligent Simulation to Enhance Human Performance in Aircraft Maintenance*, Proceedings of the International Conference on Aging Aircraft and Structural Airworthiness, 1991, pp. 305–311.

251. Johnson, W. B., Shepherd, W. T., *Impact of Human Factors Research on Commercial Aircraft Maintenance and Inspection*, Proceedings of the International Air Safety Seminar, 1993, pp. 187–199.

252. Johnson, W. B., Shepherd, W. T., *Human Factors in Aviation Maintenance: Research and Development in the USA*, Proceedings of the ICAO Flight Safety and Human Factors Seminar, 1991, pp, B.192–B.228.

253. Jones, J. A., Widjaja, T. K., *Electronic Human Factors Guide for Aviation Maintenance*, Proceedings of the Human Factors and Ergonomics Society Annual Meeting, 1995, pp. 71–74.

254. Joshi, V. V., Kaufman, L. M., Giras, T. C., *Human Behavior Modeling in Train Control Systems*, Proceedings of the Annual Reliability and Maintainability Symposium, 2001, pp. 183–188.

255. Kamiyama, M., Furukawa, A., Yoshimura, A., The Effect of Shifting Errors When Correcting Track Irregularities with a Heavy Tamping Machine, *Advances in Transport*, Vol. 7, 2000, pp. 95–104.

256. Kania, J., *Panel Presentation on Airline Maintenance Human Factors*, Proceedings of the 10th FAA Meeting on Human Factors in Aircraft, FAA/AAM Human Factors in Aviation Maintenance and Inspection Research Phase Reports (1991–1999), Brussels, Belgium, 1997.

257. Kanki, B., *Managing Procedural Error in Maintenance*, Proceedings of the Flight Safety Foundation Annual International Air Safety Seminar, 2005, pp. 233–244.

258. Kantowitz, B. H., Hanowski, R. J., Kantowitz, S. C., Driver Acceptance of Unreliable Traffic Information in Familiar and Unfamiliar Settings, *Human Factors*, Vol. 39, No. 2, 1997, pp. 164–174.

259. Karwowski, W., Parsei, H. R., Amarnath, B., Rahimi, M., A Study of Worker Intrusion in Robots Work Envelope, in *Safety, Reliability, and Human Factors in Robotic Systems*, ed. J. H. Graham, Van Nostrand Reinhold, New York, 1991, pp. 148–162.

260. Kataoka, K., Komaya, K., *Crew Operation Scheduling Based on Simulated Evolution Technique*, Proceedings of the International Conference on Computer Aided Design, Manufacture and Operation in the Railway and Other Advanced Mass Transit Systems, 1998, pp. 277–285.

261. Kaye, R., Crowley, J., *Medical Device Use-Safety: Incorporating Human Factors Engineering into Risk Management*, CDRH, Office of Health and Industry Programs, U.S. Department of Health and Human Services, Washington, D.C., 2000.

262. Kecojevic, V., Radomsky, M., The Causes and Control of Loader and Truck Related Fatalities in Surface Mining Operations, *Injury Control and Safety Promotion*, Vol. 11, No. 1, 2004, pp. 239–251.

263. Keene, S. J., *Assuring Software Safety*, Proceedings of the Annual Reliability and Maintainability Symposium, 1992, pp. 274–279.

264. Keran, C. M., Hendricks, P. A., Automation & Safety of Mobile Mining Equipment, *Engineering and Mining Journal*, Vol. 196, No. 2, 1995, pp. 30–33.

265. Kersholt, J. H., Passenier, P. O., Houttuin, K., Schuffel, H., Effect of A Priori Probability and Complexity on Decision Making in a Supervisory Control Task, *Human Factors*, Vol. 38, No. 1, 1996, pp. 65–79.

266. Khan, F. I., Haddara, M. R., Risk-Based Maintenance of Ethylene Oxide Production Facilities, *Journal of Hazardous Materials*, Vol. 108, No. 3, 2004, pp. 147–159.

267. Kioka, K., Shigemori, M., *Study on Validity of Psychological Aptitude Tests for Train Operation Divisions: A Study on Validity of Intelligence Test Pass or Failure Criterion Adopted in Japanese Railway Industry*, Quarterly Report of RTRI (Railway Technical Research Institute of Japan), Vol. 43, No. 2, 2002, pp. 63–66.

268. Kirwan, B., The Role of the Controller in the Accelerating Industry of Air Traffic Management, *Safety Science*, Vol. 37, No. 2–3, 2001, pp. 151–185.

269. Kitajima, H., Numata, N., Yamamoto, K., Goi, Y., *Prediction of Automobile Driver Sleepiness* (1st Report, Rating of Sleepiness Based on Facial Expression and Examination of Effective Predictor Indexes of Sleepiness), Nippon Kikai Gakkai Ronbunshu, C Hen/Transactions of the Japan Society of Mechanical Engineers, Part C, Vol. 63, No. 613, 1997, pp. 3059–3066.

270. Kizil, M. S., Peterson, J., English, W., The Effect of Coal Particle Size on Colorimetric Analysis of Roadway Dust, *Journal of Loss Prevention in the Process Industries*, Vol. 14, No. 5, 2001, pp. 387–394.

271. Knee, H. E., *The Maintenance Personnel Performance Simulation (MAPPS) Model: A Human Reliability Analysis Tool*, Proceedings of the International Conference on Nuclear Power Plant Aging, Availability Factor and Reliability Analysis, 1985, pp. 77–80.

272. Knox, C. E., Scanlon, C. H., Flight Tests Using Data Link for Air Traffic Control and Weather Information Exchange, *SAE (Society of Automotive Engineers) Transactions*, Vol. 99, 1990, pp. 1683–1688.

273. Kobylinski, L. K., *Rational Approach to Ship Safety Requirements*, Proceedings of the International Conference on Marine Technology, 1997, pp. 3–13.

274. Kock, A., Oberholzer, J. W., *The Development and Application of Electronic Technology to Increase Health, Safety and Productivity in the South African Coal Mining Industry*, Proceedings of the 13th IEEE Industry Applications Society Annual Meeting, 1995, pp. 2017–2022.

275. Koppa, R. J., Fambro, D. B., Zimmer, R. A., Measuring Driver Performance in Braking Maneuvers, *Transportation Research Record*, No. 1550, 1996, pp. 8–15.

276. Kovari, B., Air Crew Training, Human Factors and Reorganizing in Case of Irregularities, *Periodica Polytechnica Transportation Engineering*, Vol. 33, No. 1–2, 2005, pp. 77–88.

277. Kraft, E. R., A Hump Sequencing Algorithm for Real Time Management of Train Connection Reliability, *Journal of the Transportation Research Forum*, Vol. 39, No. 4, 2000, pp. 95–115.

278. Kraiss, K., Hamacher, N., Concepts of User Centered Automation, *Aerospace Science and Technology*, Vol. 5, No. 8, 2001, pp. 505–510.

279. Kraus, D. C., Gramopadhye, A. K., Effect of Team Training on Aircraft Maintenance Technicians: Computer-Based Training versus Instructor-Based Training, *International Journal of Industrial Ergonomics*, Vol. 27, No. 3, 2001, pp. 141–157.

280. Krauss, G. R., Cardo, A., *Safety of Life at Sea: Lessons Learned from the Analysis of Casualties Involving Ferries*, Proceedings of the International Offshore and Polar Engineering Conference, Vol. 3, 1997, pp. 484–491.

281. Kuenzi, J. K., Nelson, B. C., *Mobile Mine Equipment Maintenance Safety: A Review of U.S. Bureau of Mines Research*, Bureau of Mines, U.S. Department of the Interior, Washington, D.C., 1995.

282. Kwitowski, A. J., Brautigam, A. L., Monaghan, W. D., Teleoperated Continuous Mining Machine for Improved Safety, *Mining Engineering*, Vol. 47, No. 8, 1995, pp. 753–759.

283. Lamonde, F., Safety Improvement in Railways: Which Criteria for Coordination at a Distance Design? *International Journal of Industrial Ergonomics*, Vol. 17, No. 6, 1996, pp. 481–497.

284. Lamonde, F., Safety Improvement in Railways: Which Criteria for Coordination at a Distance Design? *International Journal of Industrial Ergonomics*, Vol. 17, No. 6, 1996, pp. 481–497.

285. Larue, C., Cohen, H. H., *Consumer Perception of Light Truck Safety*, Proceedings of the Human Factors Society 34th Annual Meeting, 1990, pp. 589–590.

286. Latorella, K. A., *Investigating Interruptions: An Example from the Flight Deck*, Proceedings of the Human Factors and Ergonomics Society Annual Meeting, Vol. 1, 1996, pp. 249–254.

287. Latorella, K. A., Drury, C. G., *Human Reliability in Aircraft Inspection, In Human Factors in Aviation Maintenance Phase II: Progress Report*, Report No. DOT/FAA/AM-93/5, Office of Aviation Medicine, Federal Aviation Administration, Washington, D.C., 1993, pp. 63–144.

288. Latorella, K. A., Drury, C. G. A., *Framework for Human Reliability in Aircraft Inspection*, Proceedings of the 7th Federal Aviation Administration Meeting on Human Factors Issues in Aircraft Maintenance and Inspection: Science, Technology, and Management: A Program Review, 1992, pp. 71–82.

289. Latorella, K. A., Prabhu, P. V., Review of Human Error in Aviation Maintenance and Inspection, *International Journal of Industrial Ergonomics*, Vol. 26, No. 2, 2000, pp. 133–161.

290. Lauber, J. K., *Contribution of Human Factors Engineering to Safety*, Proceedings of the International Air Safety Seminar, 1993, pp. 77–88.

291. Lawrence, A. C., Human Error as a Cause of Accidents in Gold Mining, *Journal of Safety Research*, Vol. 6, No. 2, 1974, pp. 78–88.

292. Layton, C. F., Shepherd, W. T., Johnson, W. B., Norton, J. E., *Enhancing Human Reliability with Integrated Information Systems for Aviation Maintenance*, Proceedings of the Annual Reliability and Maintainability Symposium, 1993, pp. 498–502.

293. Layton, C. F., Shepherd, W. T., Johnson, W. B., *Human Factors and Aircraft Maintenance*, Proceedings of the International Air Transport Association 22nd Technical Conference on Human Factors in Maintenance, 1993, pp. 143–154.

294. Le Cocq, A. D., Application of Human Factors Engineering in Medical Product Design, *Journal of Clinical Engineering*, Vol. 12, No. 4, 1987, pp. 271–277.

295. Lee, J. D., Sanquist, T. F., Augmenting the Operator Function Model with Cognitive Operations: Assessing the Cognitive Demands of Technological Innovation in Ship Navigation, *IEEE Transactions on Systems, Man, and Cybernetics: Part A: Systems and Humans*, Vol. 30, No. 3, 2000, pp. 273–285.

296. Lee, J. W., Oh, I. S., Lee, H. C., Lee, Y. H., Sim, B. S., *Human Factors Research in KAERI for Nuclear Power Plants*, Proceedings of the IEEE Sixth Annual Human Factors Meeting, 1997, pp. 13/11–13/16.

297. Lenior, T. M. J., Analyses of Cognitive Processes in Train Traffic Control, *Ergonomics*, Vol. 36, 1993, pp. 1361–1368.

298. Lerner, N., Steinberg, G., Huey, R., Hanscom, F., *Driver Misperception of Maneuver Opportunities and Requirements*, Proceedings of the XIVth Triennial Congress

of the International Ergonomics Association and 44th Annual Meeting of the Human Factors and Ergonomics Society, 2000, pp. 255–258.

299. Leveson, N. G., Harvey, P. R., Analyzing Software Safety, *IEEE Transactions on Software Engineering*, Vol. 9, No. 5, 1983, pp. 569–579.

300. Leveson, N. G., Shimeall, T. J., Safety Verification of ADA Programs Using Software Fault Trees, *IEEE Software*, July 1991, pp. 48–59.

301. Leveson, N. G., Software Safety in Computer-Controlled Systems, *IEEE Computer*, February 1984, pp. 48–55.

302. Leveson, N. G., Software Safety: Why, What, and How, *Computing Surveys*, Vol. 18, No. 2, 1986, pp. 125–163.

303. Levin, M., *Human Factors in Medical Devices: A Clear and Present Danger*, Proceedings of the First Symposium on Human Factors in Medical Devices, 1989, pp. 28–29.

304. Levkoff, B., Increasing Safety in Medical Device Software, *Medical Device & Diagnostic Industry Magazine*, Vol. 18, No. 9, 1996, pp. 92–97.

305. Li, D., Tang, W., Zhang, S., *Hybrid Event Tree Analysis of Ship Grounding Probability*, Proceedings of the International Conference on Offshore Mechanics and Arctic Engineering—OMAE, Vol. 2, 2003, pp. 345–349.

306. Li, D., Tang, W., Zhang, S., Hybrid Event Tree Calculation of Ship Grounding Probability Caused by Piloting Failure, *Shanghai Jiaotong Daxue Xuebao Journal*, Shanghai Jiaotong University, Vol. 37, No. 8, 2003, pp. 1146–1150.

307. Lin, L. J., Cohen, H. H., Accidents in the Trucking Industry, *International Journal of Industrial Ergonomics*, Vol. 20, 1997, pp. 287–300.

308. Lipowczan, A., *Increasing the Reliability and Safety of Mining Machines by Application of the Vibration Diagnostic (Experiences and Results)*, Proceedings of the International Conference on Reliability, Production, and Control in Coal Mines, 1991, pp. 155–163.

309. Litchfield, et. al., *Practical Ignition Problems Related to Intrinsic Safety in Mine Equipment: Four Short-Term Studies*, Report to the Bureau of Mines, U.S. Department of the Interior, Washington, D.C., 1980.

310. Lobb, B., Harre, N., Suddendorf, T., An Evaluation of a Suburban Railway Pedestrian Crossing Safety Programme, *Accident Analysis and Prevention*, Vol. 33, No. 2, 2001, pp. 157–165.

311. Lourens, P. F., Theoretical Perspectives on Error Analysis and Traffic Behaviour, *Ergonomics*, Vol. 33, No. 10–11, 1990, pp. 1251–1263.

312. Lucas, D., Safe People in Safe Railways, *IEE Colloquium* (Digest), No. 49, 2000, pp. 3/1–3/2.

313. Lyon, E., Miners' Electric Safety Lamps, *Electrical Review*, Vol. 98, No. 2510, 1926, pp. 9–10.

314. MacGregor, C., Hopfl, H. D., *Integrating Safety and Systems: The Implications for Organizational Learning*, Proceedings of the International Air Safety Seminar, 1992, pp. 304–311.

315. Maddox, M. E., Designing Medical Devices to Minimize Human Error, *Medical Device & Diagnostic Magazine*, Vol. 19, No. 5, 1997, pp. 166–180.

316. Maddox, M. E., *Introducing a Practical Human Factors Guide into the Aviation Maintenance Environment*, Proceedings of the Human Factors and Ergonomics Society 38th Annual Meeting, 1994, pp. 101–105.

317. Majoros, A. E., Human *Performance in Aircraft Maintenance: The Role of Aircraft Design*, Proceedings of the Meeting on Human Factors Issues in Aircraft Maintenance and Inspection, 1989, pp. A25–A32.

318. Majos, K., Communication and Operational Failures in the Cockpit, *Human Factors and Aerospace Safety*, Vol. 1, No. 4, 2001, p. 323–340.
319. Majumdar, A., Ochieng, W. Y., Nalder, P., Trend Analysis of Controller-Caused Airspace Incidents in New Zealand, 1994–2002, Transportation Research Record, No. 1888, 2004, pp. 22–33.
320. Majumdar, A., Ochieng, W. Y., A Trend Analysis of Air Traffic Occurrences in the UK Airspace, *Journal of Navigation*, Vol. 56, No. 2, 2003, pp. 211–229.
321. Majumdar, A., Ochleng, W. Y., Nalder, P., Airspace Safety in New Zealand: A Causal Analysis of Controller Caused Airspace Incidents between 1994–2002, *The Aeronautical Journal*, Vol. 108, May 2004, pp. 225–236.
322. Malavasi, G., Ricci, S., Simulation of Stochastic Elements in Railway Systems Using Self-Learning Processes, *European Journal of Operational Research*, Vol. 131, No. 2, 2001, pp. 262–272.
323. Malone, T. B., Rousseau, G. K., Malone, J. T., Enhancement of Human Reliability in Port and Shipping Operations, *Water Studies*, Vol. 9, 2000, pp. 101–111.
324. Marshall, K. L., Mine Safety as Affected by Electrification, *Coal Mining*, Vol. 5, No. 3, 1928, pp .79–80.
325. Masson, M., Koning, Y., How to Manage Human Error in Aviation Maintenance? The Example of a Jar 66-HF Education and Training Programme, *Cognition, Technology & Work*, Vol. 3, No. 4, 2001, pp. 189–204.
326. Mathews, H. W. J., *Global Outlook of Safety and Security Systems in Passenger Cars and Light Trucks*, Proceedings of the Society of Automotive Engineers Conference, 1992, pp. 71–93.
327. Matthews, R. A., Partnerships Improve Rail's Safety Record, *Railway Age*, Vol. 202, No. 3, 2001, p. 14.
328. Mayfield, T. F., Role of Human Factors Engineering in Designing for Operator Training, American Society of Mechanical Engineers *Publications on Safety Engineering and Risk Analysis* (SERA), Vol. 1, 1994, pp. 63–68.
329. Mazzeo, P. L., Nitti, M., Stella, E., Ancona, N., Distante, A., *An Automatic Inspection System for the Hexagonal Headed Bolts Detection in Railway Maintenance*, Proceedings of the IEEE Conference on Intelligent Transportation Systems, 2004, pp. 417–422.
330. McDonald, W. A., Hoffmann, E. R., Driver's Awareness of Traffic Sign Information, *Ergonomics*, Vol. 34, 1991, pp. 585–612.
331. McIvor, R. A., Mine Shaft Conveyance Safety Mechanism: Free-Fall Testing, *CIM Bulletin*, Vol. 89, No. 1004, 1996, pp. 47–50.
332. McLeod, R. W., Walker, G. H., Moray, N., Analysing and Modelling Train Driver Performance, *Applied Ergonomics*, Vol. 36, No. 6, 2005, pp. 671–680.
333. McSweeney, K. P., Baker, C. C., McCafferty, D. B., *Revision of the American Bureau of Shipping Guidance Notes on the Application of Ergonomics to Marine Systems: A Status Report*, Proceedings of the Annual Offshore Technology Conference, 2002, pp. 2577–2581.
334. Mead, D., *Human Factors Evaluation of Medical Devices*, Proceedings of the First Symposium on Human Factors in Medical Devices, 1989, pp. 17–18.
335. Meadow, L., Los Angeles Metro Blue Lime Light Rail Safety Issues, *Transportation Research Record*, No. 1433, 1994, pp. 123–133.
336. Meadows, S., *Human Factors Issues with Home Care Devices*, Proceedings of the First Symposium on Human Factors in Medical Devices, 1989, pp. 37–38.
337. Meelot, M., *Human Factor in Maintenance Activities in Operation*, Proceedings of the 10th International Conference on Power Stations, 1989, pp. 82.1–82.4.

338. Mendis, K. S., Software Safety and Its Relation to Software Quality Assurance, in *Handbook of Software Quality Assurance*, eds. G. G. Schulmeyer and J. I. McManus, Prentice Hall, Upper Saddle River, NJ, 1999, pp. 669–679.

339. Metzger, U., Parasuraman, R., Automation in Future Air Traffic Management: Effects of Decision Aid Reliability on Controller Performance and Mental Workload, *Human Factors*, Vol. 47, No. 1, 2005, pp. 35–49.

340. Meulen, M. V. D., *Definitions for Hardware and Software Safety Engineers*, Springer-Verlag, London, 2000.

341. Meyer, J. L., Some Instrument Induced Errors in the Electrocardiogram, *Journal of the American Medical Association (JAMA)*, Vol. 201, 1967, pp. 351–358.

342. Mitchell, J. S., Nuts and Bolts of Rail Safety, *Professional Engineering*, Vol. 15, No. 16, 2002, pp. 1.

343. Mjos, K., *Human Error Flight Operations*, PhD dissertation, 2002. Available from the Department of Psychology, Norwegian University of Science and Technology, Trondheim, Norway.

344. Modugno, F., Leveson, N. G., Reese, J. D., Partridge, K., Sandys, S., *Creating and Analyzing Requirement Specifications of Joint Human-Computer Controllers for Safety-Critical Systems*, Proceedings of the Annual Symposium on Human Interaction with Complex Systems, 1996, pp. 46–53.

345. Mojdehbakhsh, R., Tsai, W. T., Kirani, S., Elliott, L., Retrofitting Software Safety in an Implantable Medical Device, *IEEE Software*, No. 1, Jan. 1994, pp. 41–50.

346. Mollard, R., Coblentz, A., Cabon, P., *Vigilance in Transport Operations. Field Studies in Air Transport and Railways*, Proceedings of the Human Factors Society Annual Meeting, 1990, pp. 1062–1066.

347. Moran, J. T., *Human Factors in Aircraft Maintenance and Inspection, Rotorcraft Maintenance and Inspection*, Proceedings of the Meeting on Human Factors Issues in Aircraft Maintenance and Inspection, 1989, pp. A42–A44.

348. Moray, N., Designing for Transportation Safety in the Light of Perception, Attention, and Mental Models, *Ergonomics*, Vol. 33, No. 10–11, 1990, pp. 1201–1213.

349. Morrell, H. W., European Standards: Mining Machinery Safety, *Mining Technology*, Vol. 74, No. 851, 1992, pp. 13–14.

350. Mosier, K. L., Palmer, E. A., Degani, A., *Electronic Checklists. Implications for Decision Making*, Proceedings of the Human Factors Society Annual Conference, 1992, pp. 7–11.

351. Mount, F. E., Human Factor in Aerospace Maintenance, *Aerospace America*, Vol. 31, No. 10, 1993, pp. 1–9.

352. Nagamachi, M., Human Factors in Industrial Robots: Robot Safety Management in Japan, *Applied Ergonomics*, Vol. 17, No. 1, 1986, pp. 9–18.

353. Nagamachi, M., Ten Fatal Accidents Due to Robots in Japan, in *Ergonomics of Hybrid Automated Systems*, eds. W. Karwowski, et al., Elsevier, Amsterdam, 1988, pp. 391–396.

354. Nelson, R. H., Safety Is Good Business, *Railway Age*, Vol. 205, No. 6, 2004, p. 10.

355. Nelson, W. R., *Integrated Design Environment for Human Performance and Human Reliability Analysis*, Proceedings of the IEEE Conference on Human Factors and Power Plants, 1997, pp. 8.7–8.11.

356. Nelson, W. R., *Structured Methods for Identifying and Correcting Potential Human Errors in Aviation Operations*, Proceedings of the IEEE International Conference on Systems, Man and Cybernetics, Vol. 4, 1997, pp. 3132–3136.

357. Nibbering, J. J. W., Structural Safety and Fatigue of Ships, *International Shipbuilding Progress*, Vol. 39, No. 420, 1992, pp. 61–98.

358. Nicolaisen, P., *Ways of Improving Industrial Safety for the Programming of Industrial Robots*, Proceedings of the 3rd International Conference on Human Factors in Manufacturing, November 1986, pp. 263–276.

359. Niu, X., Huang, X., Zhao, Z., Zhang, Y., Huang, C., Cui, L., *The Design and Evaluation of a Wireless Sensor Network for Mine Safety Monitoring*, Proceedings of the IEEE Global Telecommunications Conference, 2007, pp. 1291–1295.

360. Nobel, J. J., *Human Factors Design of Medical Devices: The Current Challenge*, Proceedings of the First Symposium on Human Factors in Medical Devices, 1989, pp. 1–5.

361. Nobel, J. L., Medical Device Failures and Adverse Effects, *Pediatric Emergency Care*, Vol. 7, 1991, pp. 120–123.

362. Novak, M., Problems of Attention Decreases of Human System Operators, *Neural Network World*, Vol. 14, No. 3–4, 2004, pp. 291–301.

363. Novak, M., Votruba, Z., Challenge of Human Factor Influence for Car Safety, *Neural Network World*, Vol. 14, No. 1, 2004, pp. 37–41.

364. Novak, M., Votruba, Z., Faber, J., Impacts of Driver Attention Failures on Transport Reliability and Safety and Possibilities of its Minimizing, *Neural Network World*, Vol. 14, No. 1, 2004, pp. 49–65.

365. NSS 1740.13 Interim, Software Safety Standard, National Aeronautics and Space Administration, Washington, D.C., 1994.

366. Nunn, R., Witt, S. A., *Influence of Human Factors on the Safety of Aircraft Maintenance*, Proceedings of the International Air Safety Seminar, 1997, pp. 211–221.

367. O'Connor, S. L., Bacchi, M., *A Preliminary Taxonomy for Human Error Analysis in Civil Aircraft Maintenance Operations*, Proceedings of the Ninth International Symposium on Aviation Psychology, 1997, pp. 1008–1013.

368. Ogle, J., Guensler, R., Bachman, W., Koutsak, M., Wolf, J., Accuracy of Global Positioning System for Determining Driver Performance Parameters, *Transportation Research Record*, No. 1818, 2002, pp. 12–24.

369. Olivier, D. P., Engineering Process Improvement through Error Analysis, *Medical Device & Diagnostic Industry Magazine*, Vol. 21, No. 3, 1999, pp. 130–136.

370. Orasanu, J., Fischer, U., McDonnell, L. K., Davison, J., Haars, K. E., Villeda, E., VanAken, C., *How Do Flight Crews Detect and Prevent Errors? Findings from a Flight Simulation Study*, Proceedings of the Human Factors and Ergonomics Society Annual Meeting, Vol. 1, 1998, pp. 191–195.

371. Orlady, H. W., Orlady, L. M., Human Factors in Multi-Crew Flight Operations, *Aeronautical Journal*, Vol. 106, 2002, pp. 321–324.

372. Panzera, V. M., *Managing Operational Safety in All Phases of the Life Cycle of Railway Operations*, Proceedings of the 26th Annual Southern African Transport Conference, 2007, pp. 801–811.

373. Parasuraman, R., Hancock, P. A., Olofinboba, O., Alarm Effectiveness in Driver-Centered Collision-Warning Systems, *Ergonomics*, Vol. 40, No. 3, 1997, pp. 390–399.

374. Parker, J. F., *A Human Factors Guide for Aviation Maintenance*, Proceedings of the Federal Aviation Administration Meeting on Human Factors Issues in Aircraft Maintenance and Inspection: Science, Technology, and Management: A Program Review, 1992, pp. 207–220.

375. Parker, J. F. J., *Human Factors Guide for Aviation Maintenance*, Proceedings of the Human Factors and Ergonomics Society Annual Meeting, Vol. 1, 1993, pp. 30–35.

376. Patton, P. W., Stewart, B. M., Clark, C. C., *Reducing Materials Handling Injuries in Underground Mines*, Proceedings of the 32nd Institute on Mining Health, Safety, and Research, Salt Lake City, Utah, Aug. 5–7, 2001, pp. 1–14.

377. Pauzie, A., *Human Interface of In-Vehicle Information Systems*, Proceedings of the Conference on Vehicle Navigation and Information Systems, 1994, pp. 6–11.

378. Peacock, T., Developing Safety and Operating Standards for Rail Transit: Online on Time, and Track, *TR News*, No. 215, 2001, pp. 3–5.

379. Pekka, P., Kari, L., Lasse, R., *Study on Human Errors Related to NPP Maintenance Activities*, Proceedings of the IEEE Conference on Human Factors and Power Plants, 1997, pp. 12/23–12/28.

380. Polet, P., Vanderhaegen, F., Millot, P., *Analysis of Intentional Human Deviated Behaviour: An Experimental Study*, Proceedings of the IEEE International Conference on Systems, Man and Cybernetics, Vol. 3, 2004, pp. 2605–2610.

381. Pontt, J., Rodriguez, J., Dixon, J., *Safety, Reliability, and Economics in Mining Systems*, Proceedings of the 41st IEEE Industry Applications Conference, 2006, pp. 1–5.

382. Pyy, P., Laakso, K., Reiman, L., *A Study on Human Errors Related to NPP Maintenance Activities*, Proceedings of the IEEE Conference on Human Factors and Power Plants, 1997, pp. 12/23–28.

383. Rahimi, M., System Safety for Robots: An Energy Barrier Analysis, *Journal of Occupational Accidents*, Vol. 8, 1984, pp. 127–138.

384. Ramdass, R., *Maintenance Error Management the Next Step at Continental Airlines*, Proceedings of the Flight Safety Foundation Annual International Air Safety Seminar, 2005, pp. 115–124.

385. Randolph, R. F., Boldt, C. M. K., *Safety Analysis of Surface Haulage Accidents*, Proceedings of the 27th Annual Institute on Mining Health, Safety, and Research, 1996, pp. 29–38.

386. Ranney, T. A., Mazzae, E. N., Garrott, W. R., Barickman, F. S., *Development of a Test Protocol to Demonstrate the Effects of Secondary Tasks on Closed-Course Driving Performance*, Proceedings of the Human Factors and Ergonomics Society Annual Conference 2001, pp. 1581–1585.

387. Rao, A., Tsai, T., Safety Standards for High-Speed Rail Transportation, *Transportation Research Record*, No. 1995, 2007, pp. 35–42.

388. Rau, G., Tripsel, S., Ergonomic Design Aspects in Interaction between Man and Technical Systems in Medicine, *Medical Program Technology*, Vol. 9, 1982, pp. 153–159.

389. Reason, J., Maddox, M. E., Human Error, in *Human Factors Guide for Aviation Maintenance*, Report to the Office of Aviation Medicine, Federal Aviation Administration, Washington, D.C., 1996, pp. 14/1–14/45.

390. Reason, J., Maintenance-Related Errors: The Biggest Threat to Aviation Safety after Gravity, *Aviation Safety*, 1997, pp. 465–470.

391. Regunath, S., Raina, S., Gramopadhye, A K., *Use of HTA in Establishing Training Content for Aircraft Inspection*, Proceedings of the IIE Annual Conference, 2004, pp. 2279–2282.

392. Reid, W. S., *Safety in Perspective, for Autonomous Off Road Equipment (AORE)*, Proceedings of the ASABE Annual International Meeting, 2004, pp. 1141–1146.

393. Reinach, S., Viale, A., Application of a Human Error Framework to Conduct Train Accident/Incident Investigations, *Accident Analysis and Prevention*, Vol. 38, 2006, pp. 396–406.

394. Reynolds, R. L., History of Coal Mine Electrical Fatalities since 1970, *IEEE Transactions on Industry Applications*, Vol. 21, No. 6, 1985, pp. 1538–1544.

395. Ricci, S., Tecnologia e Comportamenti Umani Nella Sicurezza Della Circolazione Ferroviaria, (Technology and Human Behavior in Railway Traffic Safety), *Ingegneria Ferroviaria*, Vol. 56, No. 5, 2001, pp. 227–232.

396. Richards, P. G., The Perceived Gap between Need (Ed) and Mandated Training "Mind the Gap," *Aeronautical Journal*, Vol. 106, 2002, pp. 427–430.

397. Rognin, L., Salembier, P., Zouinar, M., Cooperation, Reliability of Socio-Technical Systems and Allocation of Function, *International Journal of Human Computer Studies*, Vol. 52, No. 2, 2000, pp. 357–379.

398. Rudd, D., Our Chance to Put Rail Safety First, *Professional Engineering*, Vol. 14, No. 19, 2001, pp. 17–19.

399. Ruff, T. M., Hession-Kunz, D., Application of Radio-Frequency Identification Systems to Collision Avoidance in Metal/Nonmetal Mines, *IEEE Transactions on Industry Applications*, Vol. 37, No. 1, 2001, pp. 112–116.

400. Ruff, T. M., Holden, T. P., Preventing Collisions Involving Surface Mining Equipment: A GPS-based Approach, *Journal of Safety Research*, Vol. 34, No. 2, 2003, pp. 175–181.

401. Rumar, K., Basic Driver Error. Late Detection, *Ergonomics*, Vol. 33, No. 10–11, 1990, pp. 1281.

402. Rushworth, A. M., Reducing Accident Potential by Improving the Ergonomics and Safety of Locomotive and FSV Driver Cabs by Retrofit, *Mining Technology*, Vol. 78, No. 898, 1996, pp. 153–159.

403. Russell, J. W., Robot Safety Considerations: A Checklist, *Professional Safety*, December 1983, pp. 36–37.

404. Russell, S. G., *The Factors Influencing Human Errors in Military Aircraft Maintenance*, Proceedings of the International Conference on Human Interfaces in Control Room, 1999, pp. 263–269.

405. Saccomanno, F. F., Craig, L., Shortreed, J. H., Truck Safety Issues and Recommendations of the Conference on Truck Safety: Perceptions and Reality, *Canadian Journal of Civil Engineering*, Vol. 24, No. 2, 1997, pp. 357–369.

406. Sacks, H. K., Cawley, J. C., Homce, G., Yenchek, M., *Feasibility Study to Reduce Injuries and Fatalities Caused by Contact of Cranes, Drill Rigs, and Haul Trucks with High Tension Lines*, Proceedings of the IEEE Industry Applications Society Annual Meeting, Vol. 1, 1999, pp. 240–246.

407. Sadasivan, S., Greenstein, J. S., Gramopadhye, A. K., Duchowski, A. T., *Use of Eye Movements as Feedforward Training for a Synthetic Aircraft Inspection Task*, Proceedings of the Conference on Human Factors in Computing Systems, 2005, pp. 141–149.

408. Sadasivan, S., Nalanagula, D., Greenstein, J., Gramopadhye, A., Duchowski, A., *Training Novice Inspectors to Adopt an Expert's Search Strategy*, Proceedings of the IIE Annual Conference, 2004, pp. 2257–2262.

409. *Safety in the Use and Maintenance of Large Mobile Surface Mining Equipment*, Report, Bureau of Mines, United States Department of the Interior, Washington, D.C., 1983.

410. Sammarco, J. J., Kohler, J. L., Novak, T., Morley, L.A., *Safety Issues and the Use of Software-Controlled Equipment in the Mining Industry*, Proceedings of the 32nd IEEE Industry Applications Annual Conference, 1997, pp. 2084–2090.

411. Sammarco, J. J., *Addressing the Safety of Programmable Electronic Mining Systems: Lessons Learned*, Proceedings of the IEEE 37th Industry Applications Society Annual Meeting, 2002, pp. 692–698.

412. Sammarco, J. J., *Programmable Electronic and Hardwired Emergency Shutdown Systems: A Quantified Safety Analysis*, Proceedings of the 40th IEEE Industry Applications Society Annual Meeting, 2005, pp. 210–217.

413. Sanderson, L. M., Collins, J. N., McGlothlin, J. D., Robot-Related Fatality Involving a U.S. Manufacturing Plant Employee: Case Report and Recommendations, *Journal of Occupational Accidents*, Vol. 8, 1986, pp. 13–23.

414. Sanquist, T. F., *Human Factors in Maritime Applications: A New Opportunity for Multi-Modal Transportation Research*, Proceedings of the Human Factors Society Annual Meeting, Vol. 2, 1992, pp. 1123–1127.

415. Sanquist, T. F., Lee, J. D., McCallum, M. C., *Methods for Assessing Training and Qualification Needs for Automated Ships*, Proceedings of the Human Factors and Ergonomics Society Annual Meeting, Vol. 2, 1995, pp. 1263–1267.

416. Santel, C., Trautmann, C., Liu, W., *The Integration of a Formal Safety Analysis into the Software Engineering Process: An Example from the Pacemaker Industry*, Proceedings of the Symposium on the Engineering of Computer Based-Medical Systems, 1988, pp. 52–154.

417. Sasou, K., Reason, J., Team Errors: Definition and Taxonomy, *Reliability Engineering and System Safety*, Vol. 65, No. 1, 1999, pp. 1–9.

418. Schmid, F., Organisational Ergonomics: A Case Study from the Railway Systems Area, *IEE Conference Publication*, No. 481, 2001, pp. 261–270.

419. Schmid, F., Collis, L. M., Human Centred Design Principles, *IEE Conference Publication*, No. 463, 1999, pp. 37–43.

420. Schmidt, R. A., Young, D. E., Ayres, T. J., Wong, J. R., *Pedal Misapplications: Their Frequency and Variety Revealed through Police Accident Reports*, Proceedings of the Human Factors and Ergonomics Society Annual Conference, Vol. 2, 1997, pp. 1023–1027.

421. Schreiber, P., *Human Factors Issues with Anesthesia Devices*, Proceedings of the First Symposium on Human Factors in Medical Devices, 1989, pp. 32–36.

422. Scott, A., Killing Off Errors, *Mine and Quarry*, Vol. 24, No. 9, 1995, pp. 14–18.

423. Seiff, H. E., Status Report on Large-Truck Safety, *Transportation Quarterly*, Vol. 44, No. 1, 1990, pp. 37–50.

424. Seminara, J. L., *Human Factor Methods for Assessing and Enhancing Power Plant Maintainability*, Report No. EPRI-NP-2360, Electric Power Research Institute, Palo Alto, CA, 1982.

425. Seminara, J. L., Parsons, S. O., Human Factors Engineering and Power Plant Maintenance, *Maintenance Management International*, Vol. 6, No. 1, 1985, pp. 33–71.

426. Seminara, J. L., Parsons, S. O., *Human Factors Review of Power Plant Maintainability*, Report No. EPRI-NP-1567, Electric Power Research Institute, Palo Alto, CA, 1981.

427. Senders, J. W., Medical Devices, Medical Errors, and Medical Accidents, in *Human Error in Medicine*, ed. M. S. Bogner, Lawrence Erlbaum Associates Publishers, Hillsdale, NJ, 1994, pp. 159–177.

428. Shappell, S. A., Wiegmann, D. A., *A Human Error Analysis of General Aviation Controlled Flight into Terrain Accidents Occurring between 1990–1998*, Report No. DOT/FAA/AM-03/4, Office of Aerospace Medicine, Federal Aviation Administration, Washington, D.C., March 2003.

429. Shaw, R., Safety-Critical Software and Current Standards Initiative, *Computer Methods and Programs in Biomedicine*, Vol. 44, 1994, pp. 5–22.

430. Shelden, S., Belcher, S., Cockpit Traffic Displays of Tomorrow, *Ergonomics in Design*, Vol. 7, No. 3, 1999, pp. 4–9.

431. Shepherd, A., Marshall, E., Timeliness and Task Specification in Designing for Human Factors in Railway Operations, *Applied Ergonomics*, Vol. 36, No. 6, 2005, pp. 719–727.

432. Shepherd, M., *A Systems Approach to Medical Device Safety*, Monograph, Association for the Advancement of Medical Instrumentation, Arlington, VA, 1983.

433. Shepherd, M., Brown, R., Utilizing a Systems Approach to Categorize Device-Related Failures and Define User and Operator Errors, *Biomedical Instrumentation and Technology*, November/December, 1992, pp. 461–475.

434. Shepherd, W. T., *Human Factors in Aviation Maintenance: Eight Years of Evolving R&D*, Proceedings of the Ninth International Symposium on Aviation Psychology, April 2–May 1, 1997, pp. 121–130.

435. Shepherd, W. T., *A Program to Study Human Factors in Aircraft Maintenance and Inspection*, Proceedings of the Human Factors Society 34th Annual Meeting, 1990, pp. 1168–1170.

436. Shepherd, W. T., *Human Factors Challenges in Aircraft Maintenance*, Proceedings of the Human Factors Society 36th Annual Meeting, 1992, pp. 82–86.

437. Shepherd, W. T., *Human Factors in Aviation Maintenance: Program Overview*, Proceedings of the 7th Federal Aviation Administration Meeting on Human Factors Issues in Aircraft Maintenance and Inspection: Science, Technology, and Management: Program Review, 1992, pp. 7–14.

438. Shepherd, W. T., Johnson, W. B., *Human Factors in Aviation Maintenance and Inspection: Research Responding to Safety Demands of Industry*, Proceedings of the Human Factors and Ergonomics Society 39th Annual Meeting, Vol. 1, 1995, pp. 61–65.

439. Sheriff, Y. S., Software Safety Analysis: The Characteristics of Efficient Technical Walk-Throughs, *Microelectronics and Reliability*, Vol. 32, No. 3, 1992, pp. 407–414.

440. Shinomiya, A., *Recent Researches of Human Science on Railway Systems*, Quarterly Report of RTRI (Railway Technical Research Institute of Japan), Vol. 43, No. 2, 2002, pp. 54–57.

441. Shorrock, S. T., Errors of Memory in Air Traffic Control, *Safety Science*, Vol. 43, No. 8, 2005, pp. 571–588.

442. Shorrock, S. T., Kirwan, B., Development and Application of a Human Error Identification Tool for Air Traffic Control, *Applied Ergonomics*, Vol. 33, No. 4, 2002, pp. 319–336.

443. Shorrock, S. T., Kirwan, B., MacKendrick, H., Scaife, R., Foley, S., The Practical Application of Human Error Assessment in UK Air Traffic Management, *IEE Conference Publication*, No. 481, 2001, pp. 190–195.

444. Simpson, G. C., Promoting Safety Improvements via Potential Human Error Audits, *Mining Engineer*, Vol. 154, No. 395, 1994, pp. 38–42.

445. Singer, G., Starr, A., *Improvement by Regulation: Addressing Flight Crew Error/ Performance in a New Flight Deck Certification Process*, Proceedings of the Annual European Aviation Safety Seminar, 2004, pp. 83–87.

446. Siregar, M. L., Kaligis, W. K., Viewing Characteristics of Drivers, *Advances in Transport*, Vol. 8, 2001, pp. 579–587.

447. Small, D. W., Kerns, K., *Opportunities for Rapid Integration of Human Factors in Developing a Free Flight Capability*, Proceedings of the AIAA/IEEE Digital Avionics Systems Conference, 1995, pp. 468–473.

448. Smith, C. E., Peel, D., Safety Aspects of the Use of Microprocessors in Medical Equipment, *Measurement and Control*, Vol. 21, No. 9, 1988, pp. 275–276.

449. Son, K., Choi, K., Yoon, J., Human Sensibility Ergonomics Approach to Vehicle Simulator Based on Dynamics, *JSME International Journal, Series C: Mechanical Systems, Machine Elements and Manufacturing*, Vol. 47, No. 3, 2004, pp. 889–895.

450. Song, H. S., Kim, T. G., Application of Real-Time DEUS to Analysis of Safety-Critical Embedded Control Systems: Railroad Crossing Control Example, *Simulation*, Vol. 81, No. 2, 2005, pp. 119–136.

451. Sraeter, O., Kirwan, B., *Differences between Human Reliability Approaches in Nuclear and Aviation Safety*, Proceedings of the IEEE 7th Human Factors Meeting, 2002, pp. 3.34–3.39.

452. Stager, P., Hameluck, D., Ergonomics in Air Traffic Control, *Ergonomics*, Vol. 33, No. 4, 1990, pp. 493–499.

453. Stewart, B., Iverson, S., Beus, M., Safety Considerations for Transport of Ore and Waste in Underground Ore Passes, *Mining Engineering*, Vol. 51, No. 3, 1999, pp. 53–60.

454. Stormont, D., Offshore Rig Designed for Safety, *Oil and Gas Journal*, Vol. 57, No. 9, 1959, pp. 147–148, 151.

455. Strauch, B., Sandler, C. E., *Human Factors Considerations in Aviation Maintenance*, Proceedings of the Human Factors Society 28th Annual Meeting, Vol. 2, 1984, pp. 913–915.

456. Subramanian, S., Elliott, L., Vishnuvajjala, R. V., Tsai, W. T., Mojdejbakhsh, R., *Fault Mitigation in Safety-Critical Software Systems*, Proceedings of the 9th IEEE Symposium on Computer-Based Medical Systems, 1996, pp. 12–17.

457. Sugimoto, N., Subject and Problems of Robot Safety Technology, in *Occupational Health and Safety in Automation and Robotics*, ed. K. Noro, Taylor and Francis, London, 1987, pp. 175–195.

458. Sweeney, M. M., Ellingstad, V. S., Mayer, D. L., Eastwood, M. D., Weinstein, E. B., Loeb, B. S., *The Need for Sleep: Discriminating between Fatigue-Related and Nonfatigue-Related Truck Accidents*, Proceedings of the Human Factors and Ergonomics Society Annual Conference, Vol. 2, 1995, pp. 1122–1126.

459. Sykes, E. H., Mine Safety Appliances, *Canadian Mining Journal*, Vol. 48, No. 43, 1927, pp. 856–861.

460. Taylor, M., Integration of Life Safety Systems in a High-Risk Underground Environment, *Engineering Technology*, Vol. 8, No. 7, 2005, pp. 42–47.

461. Telle, B., Vanderhaegen, F., Moray, N., *Railway System Design in the Light of Human and Machine Unreliability,* Proceedings of the IEEE International Conference on Systems, Man and Cybernetics, Vol. 4, 1996, pp. 2780–2785.

462. Thompson, P. W., Safer Design of Anaesthesia Equipment, *British Journal of Anaesthesia*, Vol. 59, 1987, pp. 913–921.

463. Tien, J. C., Health and Safety Challenges for China's Mining Industry, *Mining Engineering*, Vol. 57, No. 4, 2005, pp. 15–23.

464. Tsai, T., Lamond, J. H., Liao, S. S. C., *Impact Resistance of Rail Vehicle Window Glazing and Related Safety Issues*, Proceedings of the ASME International Mechanical Engineering Congress and Exposition, 2005, pp. 143–151.

465. Tsuchiya, M., Ikeda, H., *Human Reliability Analysis of LPG Truck Loading Operation*, IFAC Symposia Series, No. 6, 1992, pp. 135–139.

466. Tsukamoto, D., Hasegawa, T., *Development of Maintenance Support System for Wayside Workers*, Quarterly Report of RTRI (Railway Technical Research Institute of Japan), Vol. 43, No. 4, 2002, pp. 175–181.

467. Turcic, P. M., *Health and Safety Implications of the Use of Diesel-Powered Equipment in Underground Coal Mines*, Proceedings of the Third Mine Ventilation Symposium, 1987, p. 390.

468. Ugajin, H., *Human Factors Approach to Railway Safety*, Quarterly Report of RTRI (Railway Technical Research Institute of Japan), Vol. 40, No. 1, 1999, pp. 5–10.

469. Ujimoto, K. V., Integrating Human Factors into the Safety Chain: A Report on International Air Transport Association's (IATA) Human Factors '98, *Canadian Aeronautics and Space Journal*, Vol. 44, No. 3, 1998, pp. 194–197.

470. Ulrich, K. T., Tuttle, T. T., Donoghue, J. P., Townsend, W. T., *Intrinsically Safer Robots*, NASA Report No. NAS 10-12178, Barrett Technology, Inc., Cambridge, MA, May 1995.

471. Urban, K., Safety in the Design of Offshore Platforms: Integrated Safety Versus Safety as an Add-On Characteristic, *Safety Science*, Vol. 45, No. 1–2, 2007, pp. 107–127.

472. Vakil, S. S., Hansman, R. J., *Predictability as a Metric of Automation Complexity*, Proceedings of the Human Factors and Ergonomics Society Annual Meeting, Vol. 1, 1997, pp. 70–74.

473. Van Elslande, P., Fleury, D., *Elderly Drivers: What Errors Do They Commit on the Road?* Proceedings of the XIVth Triennial Congress of the International Ergonomics Association and 44th Annual Meeting of the Human Factors and Ergonomics Society, 2000, pp. 259–262.

474. Vanderhaegen, F., APRECIH: A Human Unreliability Analysis Method— Application to Railway System, *Control Engineering Practice*, Vol. 7, No. 11, 1999, pp. 1395–1403.

475. Vanderhaegen, F., Non-Probabilistic Prospective and Retrospective Human Reliability Analysis Method—Application to Railway System, *Reliability Engineering and System Safety*, Vol. 71, 2001, pp. 1–13.

476. Vanderhaegen, F., Telle, B., *Consequence Analysis of Human Unreliability during Railway Traffic Control*, Proceedings of the International Conference on Computer Aided Design, Manufacture and Operation in the Railway and Other Advanced Mass Transit Systems, 1998, pp. 949–958.

477. Varma, V., Maintenance Training Reduces Human Errors, *Power Engineering*, Vol. 100, No. 8, 1996, pp. 44, 46–47.

478. Visciola, M., *Pilot Errors. Do We Know Enough?* Proceedings of the International Air Safety Seminar, 1990, pp. 11–17.

479. Vora, J., Nair, S., Gramopadhye, A. K., Melloy, B. J., Medlin, E., Duchowski, A. T., Kanki, B. G., *Using Virtual Reality Technology to Improve Aircraft Inspection Performance: Presence and Performance Measurement Studies*, Proceedings of the Human Factors and Ergonomics Society Annual Meeting, 2001, pp. 1867–1871.

480. Wade, J., In the Frame for Safer Heavy Vehicles: Heavy Vehicle Accidents in Mining, *Australian Mining*, Vol. 98, No. 9, 2006, p. 40.

481. Wagenaar, W. A., Groenewcg, J., Accidents at Sea: Multiple Causes and Impossible Consequences, *Int. J. Man-Machine Studies,* Vol. 27, 1987, pp. 587–598.

482. Wallace, D. R., Kuhn, D. R., Ippolito, L. M., *An Analysis of Selected Software Safety Standards*, Proceedings of the Seventh Annual Conference on Computer Assurance, 1992, pp. 123–136.

483. Walter, D., Competency-Based On-the-Job Training for Aviation Maintenance and Inspection: A Human Factors Approach, *International Journal of Industrial Ergonomics*, Vol. 26, No. 2, 2000, pp. 249–259.

484. Wang, J., Offshore Safety Case Approach and Formal Safety Assessment of Ships, *Journal of Safety Research*, Vol. 33, No. 1, 2002, pp. 81–115.

485. Watson, G. S., Papelis, Y. E., Chen, L. D., Transportation Safety Research Applications Utilizing High-Fidelity Driving Simulation, *Advances in Transport,* Vol. 14, 2003, pp. 193–202.

486. Wedzinga, A. A., *Revolutionary Modernization and Simplification of Air Traffic Safety Rules on Dutch Railways,* Proceedings of the 1st World Congress on Safety of Transportation, 1992, pp. 138–149.

487. Weide, P., Improving Medical Device Safety with Automated Software Testing, *Medical Device & Diagnostic Industry Magazine*, Vol. 16, No. 8, 1994, pp. 66–79.

488. Weinberg, D. I., Artley, J. A., Whalen, R. E., McIntosh, M. T., Electrical Shock Hazards in Cardiac Catheterization, *Circulation Research*, Vol. 11, 1962, pp. 1004–1011.

489. Weinger, M. B., Anesthesia Equipment and Human Error, *Journal of Clinical Monitoring & Computing,* Vol. 15, No. 5, 1999, pp. 319–323.

490. Welch, D. L., Human Error and Human Factors Engineering in Health Care, *Biomedical Instrumentation & Technology*, Vol. 31, No. 6, 1997, pp. 627–631.

491. Welch, D. L., Human Factors Analysis and Design Support in Medical Device Development, *Biomedical Instrumentation & Technology*, Vol. 32, No. 1, 1998, pp. 77–82.

492. Wenner, C. A., Drury, C. G., Analyzing Human Error in Aircraft Ground Damage Incidents, *International Journal of Industrial Ergonomics*, Vol. 26, 2000, pp. 177–199.

493. Wenner, C. L., Wenner, F., Drury, C. G., Spencer, F., *Beyond "Hits" and "Misses." Evaluating Inspection Performance of Regional Airline Inspectors,* Proceedings of the 41st Annual Human Factors and Ergonomics Society Meeting, 1997, pp. 579–583.

494. West, R., French, D., Kemp, R., Elander, J., Direct Observation of Driving, Self Reports of Driver Behavior, and Accident Involvement, *Ergonomics*, Vol. 36, No. 5, 1993, pp. 557–567.

495. Whalen, R. E., Starmer, C. F., McIntosh, H. D., Electrical Hazards Associated with Cardiac Pacemaking, *Transactions of the New York Academy of Sciences,* Vol. 111, 1964, pp. 922–931.

496. White, P., Dennis, N., Tylor, N., Analysis of Recent Trends in Bus and Coach Safety in Britain, *Safety Science*, Vol. 19, No. 2–3, 1995, pp. 99–107.

497. Wigglesworth, E. C., A Human Factors Commentary on Innovations at Railroad-Highway Grade Crossings in Australia, *Journal of Safety Research*, Vol. 32, No. 3, 2001, pp. 309–321.

498. Wiklund, M. E., Human Error Signals Opportunity for Design Improvement, *Medical Device & Diagnostic Industry Magazine*, Vol. 14, No. 2, 1992, pp. 57–61.

499. Wiklund, M. E., Making Medical Device Interfaces More User Friendly, *Medical Device & Diagnostic Industry Magazine*, Vol. 20, No. 5, 1998, pp. 177–183.

500. Wilcox, S. B., *Building Human Factors into the Design of Medical Products*, Proceedings of the First Symposium on Human Factors in Medical Devices, 1989, pp. 14–16.

501. Williams, M., Hollands, R., Schoffield, D., Denby, B., *Virtual Haulage Trucks: Improving the Safety of Surface Mines*, Proceedings of the Regional APCOM Conference, 1998, pp. 337–344.

502. Wilner, F. H., Rail Safety's Newest Face, *Railway Age*, Vol. 205, No. 10, 2004, p. 44.

503. Wilson, J. R., Norris, B. J., Rail Human Factors: Past, Present and Future, *Applied Ergonomics*, Vol. 36, No. 6, 2005, pp. 649–660.

504. Wood, B. J., Ermes, J. W., Applying Hazard Analysis to Medical Devices, Part II, *Medical Device & Diagnostic Industry Magazine*, Vol. 15, No. 3, 1993, pp. 58–64.

505. Woodward, J. B., Parsons, M. G., Troesch, A. W., Ship Operational and Safety Aspects of Ballast, Water Exchange at Sea, *Marine Technology*, Vol. 31, No. 4, 1994, pp. 315–326.

506. Wright, K., Embrey, D., Using the MARS Model for Getting at the Causes of SPADS, *Rail Professional*, 2000, pp. 6–10.

507. Wu, L., Yang, Y., Jing, G., *Application of Man-Machine-Environment System Engineering in Underground Transportation Safety*, Proceedings in Mining Science and Safety Technology Conference, 2002, pp. 514–518.

508. Xiaoli, L., *Classified Statistical Report on 152 Flight Incidents of Less Than Separation Standard Occurred in China Civil Aviation during 1990–2003*, Proceedings of the 2004 International Symposium on Safety Science and Technology, 2004, pp. 166–172.

509. Ye, L., Jiang, Y., Jiang, W., Shen, M., *Locomotive Drivers' Diatheses in China's Railways*, Proceedings of the Conference on Traffic and Transportation Studies, ICTTS, Vol. 4, 2004, pp. 392–396.

510. Yemelyanov, A. M., *Unified Modeling of Human Operator Activity in a Real-World Environment*, Proceedings of the IEEE International Conference on Systems, Man and Cybernetics, Vol. 3, 2005, pp. 2476–2481.

511. Zeng, W., *Exploration for Human Factors in the Design of Coal-Mine Safety and Rescue Devices*, Proceedings of the 7th International Conference on Computer-Aided Industrial Design and Conceptual Design, 2006, pp. 412–416.

512. Zhang, R. X., et al., Surface Mine System Simulation and Safety Risk Management, *Journal of China University of Mining and Technology*, Vol. 16, No. 4, 2006, pp. 413–415.

513. Ziskovsky, J. P., Working Safely with Industrial Robots, *Plant Engineering*, May 1984, pp. 81–85.

Index

Printed and bound by CPI Group (UK) Ltd, Croydon, CR0 4YY

23/10/2024

01778242-0010